Science and Technology Advice for Congress

Edited by
M. Granger Morgan
Jon M. Peha

RESOURCES FOR THE FUTURE WASHINGTON, DC

Printed in the United States of America

An RFF Press book
Published by Resources for the Future
1616 P Street, NW
Washington, DC 20036–1400
USA
www.rffpress.org

Library of Congress Cataloging-in-Publication Data

Science and technology advice for Congress / edited by M. Granger Morgan and Jon M. Peha
p. cm.
Includes bibliographic references and index.
ISBN 1–891853–75–9 (hardcover : alk. paper) — ISBN 1–891853–74–0 (pbk. : alk. paper)
1. Science and state—United States. 2. Technology and state—United States. 3. United States. Congress. Office of Technology Assessment. I. Morgan, M. Granger (Millett Granger), 1941– II. Peha, Jon M.
Q127 .U5S34 2003
338.973'06—dc21 2003013967

f e d c b a

The paper in this book meets the guidelines for permanence and durability of the Committee on Production Guidelines for Book Longevity of the Council on Library Resources.

This book was designed and typeset in Trump Medieval and ITC Franklin Gothic by Betsy Kulamer. It was copyedited by Paula Bérard. The cover was designed by Rosenbohm Graphic Design.

The findings, interpretations, and conclusions offered in this publication are those of the contributors and are not necessarily those of Resources for the Future, its directors, or its officers.

ISBN 1–891853–75–9 (cloth) ISBN 1–891853–74–0 (paper)

About Resources for the Future *and* RFF Press

Resources for the Future (RFF) improves environmental and natural resource policymaking worldwide through independent social science research of the highest caliber. Founded in 1952, RFF pioneered the application of economics as a tool to develop more effective policy about the use and conservation of natural resources. Its scholars continue to employ social science methods to analyze critical issues concerning pollution control, energy policy, land and water use, hazardous waste, climate change, biodiversity, and the environmental challenges of developing countries.

RFF Press supports the mission of RFF by publishing book-length works that present a broad range of approaches to the study of natural resources and the environment. Its authors and editors include RFF staff, researchers from the larger academic and policy communities, and journalists. Audiences for publications by RFF Press include all of the participants in the policymaking process—scholars, the media, advocacy groups, NGOs, professionals in business and government, and the public.

Contents

Preface vii

Contributors ix

Part I: The Issue

1. Analysis, Governance, and the Need for Better Institutional Arrangements 3
   M. Granger Morgan and Jon M. Peha

Part II: Background

2. Technical Advice for Congress: Past Trends and Present Obstacles 23
   Bruce L.R. Smith and Jeffrey K. Stine

3. The Origins, Accomplishments, and Demise of the Office of Technology Assessment. 53
   Robert M. Margolis and David H. Guston

4. Insights from the Office of Technology Assessment and Other Assessment Experiences 77
   David H. Guston

5. The European Experience 90
   Norman J. Vig

Part III: Possible Institutional Models

6. Thinking about Alternative Models 101
M. Granger Morgan and Jon M. Peha

7. An Expanded Analytical Capability in the Congressional Research Service, the General Accounting Office, or the Congressional Budget Office 106
Christopher T. Hill

8. Expanded Use of the National Academies 118
John Ahearne and Peter Blair

9. Expanding the Role of the Congressional Science and Engineering Fellowship Program 134
Albert H. Teich and Stephen J. Lita

10. A Lean, Distributed Organization To Serve Congress 145
M. Granger Morgan, Jon M. Peha, and Daniel E. Hastings

11. A Dedicated Organization in Congress 157
Gerald L. Epstein and Ashton B. Carter

12. An Independent Analysis Group That Works Exclusively for Congress, Operated by a Nongovernmental Organization 164
Caroline S. Wagner and William A. Stiles Jr.

Part IV: Moving toward Solution

13. Where Do We Go from Here? 173
M. Granger Morgan and Jon M. Peha

Appendix 1: The Technology Assessment Act of 1972. 183

Appendix 2: Details on the National Academies Complex 191

Appendix 3: An External Evaluation of the GAO's First Pilot Technology Assessment 208

Index. 229

Preface

This book had its origins in a workshop held in Washington, D.C., on June 14, 2001, "Creating Institutional Arrangements To Provide Science and Technology Advice to the U.S. Congress." After meeting over breakfast with several members of Congress on Capitol Hill, a group of more than 100 congressional staffers, policy analysts, academics, and others spent the day exploring the question "what new institutional arrangements (if any) are needed to better provide balanced, independent scientific and technical advice to Congress on large-scale questions that require foresight, analysis, and synthesis?"

House Science Chairman Sherwood Boehlert (Republican, New York), along with Representatives Vernon Ehlers (Republican, Michigan), Rush Holt (Democrat, New Jersey), and Amo Houghton (Republican, New York), participated in the breakfast and made remarks. Senators John D. Rockefeller IV (Democrat, West Virginia) and Ted Stevens (Republican, Alaska) were unable to participate because of schedule conflicts, but they sent supporting letters. In his letter, Stevens noted that "there is great need for balanced analysis to assist the Congress as it addresses complex, large-scale issues involving science and technology."

The workshop had no short-term political agenda and was not undertaken with the objective of promoting any specific institutional solution. Rather, the objective was to start a serious national discussion on this topic; in that, it was highly successful. In the weeks and months that followed, a large number of news and commentary pieces appeared in the general and science and technology specialty press, serious discussions of the issues were held among various groups in Congress, and several pieces of legislation were introduced.

Science and Technology Advice for Congress is not a proceedings of the June 2001 workshop. Rather, it is intended as another contribution to the continuing discussion that the workshop began. A number of scholars and policy experts prepared background papers for the workshop. These have been edited and expanded and merged with several newly written chapters to comprise this book.

Acknowledgements

The workshop was convened by a group of 18 leading professional societies, universities, and think tanks. We thank them for their support. They included:

- the American Association for the Advancement of Science;
- the American Association of Engineering Societies;
- the Engineering Deans Council of the American Society of Engineering Education;
- the American Society of Mechanical Engineers;
- the U.S. Public Policy Committee of the Association for Computing Machinery;
- the Department of Engineering and Public Policy of Carnegie Mellon University;
- the Center for Science, Policy and Outcomes of Columbia University;
- the Consortium of Social Science Associations;
- the School of Public Policy at George Mason University;
- the H. John Heinz III Center for Science, Economics and the Environment;
- the Kennedy School of Government at Harvard University;
- the Institute of Electrical and Electronics Engineers—USA;
- the Engineering Systems Division of Massachusetts Institute of Technology;
- Resources for the Future;
- Sigma Xi (the Scientific Research Society);
- the Society for Risk Analysis;
- the Department of Management Science and Engineering at Stanford University; and
- the College of Engineering of the University of Florida.

In addition to representatives from these organizations, and our coauthors, we owe a special debt of gratitude to Patricia Steranchak and Alexandra Carr for their assistance and support in this project.

Financial support for this workshop has come from the Heinz Endowments, the John D. and Catherine T. MacArthur Foundation, and the Department of Engineering and Public Policy at Carnegie Mellon University.

Contributors

John Ahearne is director of the Ethics Program at the Sigma Xi Center in Research Triangle Park, North Carolina and lecturer in public policy studies at Duke University. He is chair of the Board on Radioactive Waste Management in the National Research Council's Division on Earth and Life Sciences and was chairman of the Nuclear Regulatory Commission, Principal Deputy Assistant Secretary of Defense, and Deputy Assistant Secretary of Energy. He also served as vice president of Resources for the Future and executive director of Sigma Xi.

Peter Blair is executive director of the Division on Engineering and Physical Sciences of the National Research Council and adjunct professor of public policy at the University of North Carolina at Chapel Hill. He was executive director of Sigma Xi and served as energy program director and assistant director at the former U.S. Office of Technology Assessment.

Ashton B. Carter is codirector, with William J. Perry, of the Preventive Defense Project, a research collaboration between Harvard University's John F. Kennedy School of Government and Stanford University, and Ford Foundation Professor of Science and International Affairs at the Kennedy School. Carter served as Assistant Secretary of Defense for International Security Policy. He is author of *Preventive Defense: A New Security Strategy for America* (with William J. Perry).

Gerald L. Epstein is a research staff member in the Strategy, Forces, and Resources Division of the Institute for Defense Analyses. Previously, he studied science, technology, and national security issues at the U.S. Office of Technology Assessment; directed a faculty study at Harvard University's Kennedy School of Government on the relationship between military and commercial technologies; and worked at the White House Office of Science and Technology Policy. He is a coauthor of *Beyond Spinoff: Military and Commercial Technologies in a Changing World.*

David H. Guston is an associate professor in the Edward J. Bloustein School of Planning and Public Policy at Rutgers, the State University of New Jersey, where he directs the program in public policy. He is a faculty associate at Columbia University's Center for Science, Policy, and Outcomes and of the Belfer Center for Science and International Affairs at Harvard's Kennedy School of Government. He previously worked at the National Academy of Sciences and the U.S. Office of Technology Assessment. He is the author of *Between Politics and Science: Assuring the Integrity and Productivity of Research*; coauthor (with Megan Jones and Lewis M. Branscomb) of *Informed Legislatures*; and coeditor (with Kenneth Keniston) of *The Fragile Contract.* He is also the North American editor of *Science and Public Policy.*

Daniel E. Hastings is professor of aeronautics and astronautics and engineering systems and codirector of the Engineering Systems Division in the School of Engineering, and he was the director of the Technology and Policy Program at the Massachusetts Institute of Technology. As chief scientist of the Air Force, Hastings served as chief scientific adviser to the Chief of Staff and the Secretary of Defense.

Christopher T. Hill is vice provost for research and professor of public policy and technology at George Mason University; he is also president of George Mason Intellectual Properties, Inc. He has held professional positions in industry, government, and universities, including service as a senior specialist in science and technology policy at the Congressional Research Service and as a senior staff member of the U.S. Office of Technology Assessment. He is the editor with James Utterback of *Technological Innovation for a Dynamic Economy* and coauthor of *Regulation, Market Price, and Process Innovation*.

Stephen J. Lita is a project coordinator with the Directorate for Science and Policy Programs at the American Association for the Advancement of Science (AAAS), where he has been a co-organizer of the annual AAAS Colloquium on Science and Technology Policy and editor of the AAAS Science and Technology Policy Yearbook.

Robert M. Margolis is a senior energy analyst in the Washington, D.C., office of the National Renewable Energy Laboratory. His main research interests include energy technology and policy; research, development, and demonstration policy; and energy-economic-environmental modeling. Previously, he was a member of the research faculty in the Department of Engineering and Public Policy at Carnegie Mellon University and a research fellow in the Belfer Center for Science and International Affairs at Harvard's Kennedy School of Government.

M. Granger Morgan is professor and head of the Department of Engineering and Public Policy at Carnegie Mellon University, where he is also University Professor and Lord Chair Professor in Engineering; he is also a professor in the Department of Electrical and Computer Engineering and in the H. John Heinz III School of Public Policy and Management. He is a coauthor with Max Henrion of *Uncertainty: A Guide to Dealing with Uncertainty in Quantitative Risk and Policy Analysis* and with Baruch Fischhoff, Ann Bostrom, and Cynthia J. Atman of *Risk Communication: A Mental Models Approach*.

Jon M. Peha is a professor at Carnegie Mellon University in the Department of Engineering and Public Policy. He also is affiliated with the Department of Electrical and Computer Engineering and is associate director of Carnegie Mellon's Center for Wireless and Broadband Networks. Previously, Peha handled telecommunications and electronic commerce issues on legislative staff for the House Commerce Committee and for Senator Ron Wyden (Democrat, Oregon); launched a federal interagency program to assist developing countries with information infrastructure; and served on the technical staffs of SRI International, AT&T Bell Laboratories, Microsoft, and several start-up businesses.

Bruce L.R. Smith is a senior scholar at the Heyman Center for the Humanities at Columbia University. Previously, he served on the faculty of Columbia University as professor of public law and government and as a senior staff member at the Brookings Institution. Smith is the author or editor of 16 books, including *American Science Policy since World War II* and *The Advisors: Scientists in the Policy Process*.

William A. Stiles Jr. is a consultant on science policy and an adjunct faculty member at Old Dominion University. He worked for 22 years in the House of Representatives on science policy and environmental issues for the late Congressman George E. Brown Jr. (Democrat, California) and the House Science Committee, where he was legislative director. He also served as congressional board staff for the U.S. Office of Technology Assessment. He currently is a member of the Committee on Science, Engineering, and Public Policy at the American Association for the Advancement of Science.

Jeffrey K. Stine is curator of engineering and environmental history at the Smithsonian Institution's National Museum of American History. Previously, he served as an American Historical Association Congressional Fellow with the House Committee on Science and Technology, where he assisted the special Task Force on Science Policy by writing the background report, *A History of Science Policy in the United States, 1940–1985*. He is the author of *Mixing the Waters: Environment, Politics, and the Building of the Tennessee–Tombigbee Waterway* and *Twenty Years of Science in the Public Interest: A History of the Congressional Science and Engineering Fellowship Program*.

Albert H. Teich is director of science and policy programs at the American Association for the Advancement of Science (AAAS). He is the editor of several books, including *Technology and the Future*, 9th edition. He is founding codirector of the Center for Innovation Policy Research and Education in Budapest, Hungary; chair of the Advisory Board of the School of Public Policy at Georgia Institute of Technology; a member of the Board of Governors of the U.S.–Israel Binational Science Foundation; and a member of advisory boards to the Loka Institute in Amherst, Massachusetts, and Columbia University's Center for Science, Policy, and Outcomes, in Washington, D.C.

Norman J. Vig is Winifred and Atherton Bean Professor of Science, Technology, and Society, Emeritus at Carleton College, where he served as professor of political science and founding director of the Environmental and Technology Studies Program. He is coeditor of several books including *Technology and Politics*, *The Global Environment*, *Parliaments and Technology: The Development of Technology Assessment in Europe*, and *Environmental Policy: New Directions for the Twenty-First Century*, 5th edition.

Caroline S. Wagner is a research leader conducting analysis of science and technology policy issues, currently working out of RAND's offices in Leiden, The Netherlands, and at the University of Amsterdam. Past positions within RAND include deputy to the director of the Science and Technology Policy Institute, director of outreach for the institute, and policy analyst. Previously, Wagner was a professional staff member for the House of Representatives Committee on Science, Space, and Technology and an analyst for the U.S. government specializing in comparative analysis of scientific and technical capabilities worldwide.

Part I

The Issue

1

Analysis, Governance, and the Need for Better Institutional Arrangements

M. Granger Morgan and Jon M. Peha

The views of Americans about the role of expertise in governance have long been a source of disagreement and tension. According to historian Richard D. Brown (1996), the founding fathers believed that important issues of public policy should "be considered only by deliberative, representative assemblies." The idea "that private citizens would be sufficiently informed to make policy in all ... [the important areas that government must address] was never contemplated and would have seemed absurd," at least to the Federalists who carried the day in the constitutional debates. Thus, important policy decisions in the United States are made by representative legislative bodies like Congress instead of popular election because the founding fathers assumed that representatives would have greater mastery of relevant information and expertise.

At the same time, Americans have a longstanding tradition of valuing leaders who are simple, honest, and close to the people. Some voters have expressed greater interest in whether their senator knows the price of a loaf of bread than whether he or she understands the intricate details of agricultural policy. Popular books and movies such as the 1939 Frank Capra classic "Mr. Smith Goes to Washington" celebrate the simple, honorable citizen turned legislator and deride the career politician. In recent years, such views have taken practical form in successful efforts in several states to institute term limits; proposals to trim the size of congressional staff (Joint Committee on the Organization of Congress 1993), which led to actual reductions in 1995; and ridicule of executive branch bureaucratic experts. These ideas are not new. Cecilia M. Kenyon (1966) argues that similar ideas, including a desire for annual elections, mandatory rotation, and an "overt anti-intellectualism," characterized many of the arguments of the Antifederalists. She quotes, for example, a group of Pennsylvanians who in 1788 delighted in seeing "... a few country farmers and mechanics nonplus" the educated elite and rejoiced "... to see scholastic learning and erudition fly before simple reason, plain truth and common sense" (Kenyon 1966).

No matter how long they remain in office, legislators cannot become experts in every issue that comes before them. However, Brown (1996) notes that, "although no one believes it, in political rhetoric Americans pretend that they and their officials are sufficiently informed to be omnicompetent. Since the Jacksonian era, this has been one of the agreed-upon fictions of democracy. But, like many hypocrisies, this fiction encourages unrealistic expectations and disillusionment."

In his analysis of decisions made by House members in 1969, based on 222 member interviews on 15 votes, John W. Kingdon (1973) notes, "Like most busy people, congressmen have limited time to devote to any one activity and are faced with much more information than they can systematically sift and consider. Most of them, therefore, engage in an extended search for information only rarely, and then only when confronted with some unusual problem, such as an intense conflict among the factors they usually weigh, a new situation not governed by their past voting pattern, or an issue on which it is difficult to use their ideology." If there is no time to sift through the evidence and become experts, do legislators seek out the leading experts for advice on voting decisions? According to Kingdon (1973), constituency views and the views of other members are by far the most influential factors in shaping a member's decisions on how to vote.

Edward V. Schneier and Bertram Gross (1993) also observed, "Most legislative decisions are made by men and women who have little direct knowledge of the problem at issue." "The best I can do," they quote one typical member as saying, "is to devote my time to the major legislation which comes before the House and rely on others for advice in connection with less major legislation." In legislatures as in life, they argue, "informed decision can be made by uninformed human beings. Indeed they usually are, or at least they are made by persons who lack direct, personal knowledge of the relevant facts. If the sign says 'Detour bridge out ahead,' we are usually better off relying on the instructions of those who make such signs than personally testing bridges. In a complex world, it is rational and essential for each individual to remain ignorant of the details of most of the components of most of the decisions he or she must make."

Brown concludes his history of the idea of the informed citizen (1996) with the slightly different observation that "as members of a society in which we readily defer to experts every day in a multitude of occupations and professions—whether pilots, physicians, accountants, or electricians—we should also recognize that the complicated political judgements required to operate the machinery of American society with some efficiency, justice, and improvement also demand experience, learning, judgement, and character. The voice of the people, though it is only slightly informed, must be heard in the political process, but must not be the only voice. Ultimately, people who are well-informed should make policy, taking public opinion into account."

In short, Americans' views about the appropriate role of expertise in governance remain a source of disagreement and tension.

Congressional Information Needs

Congress frequently makes decisions that involve complex issues of science and technology. It must make choices about the focus and funding of federal programs to conduct or support research and development and assess the benefits and other consequences likely to result from such investments. Congress must make judgements about how available science and technology can best be brought to bear in addressing important social problems or in providing insight that can help resolve contentious policy issues. Also, it must consider the potential societal implications of current scientific and technical developments, and the regulatory and other needs that they may create, as well as emerging scientific and technical developments and the social responses that they may require. Box 1-1 provides examples of all of these.

Congress has many sources to which it can turn when it needs information. When members and committees need facts and figures or modest technical explanations, they can get assistance from the Congressional Research Service (CRS). Most staff and members have friends and colleagues to whom they turn when they have questions. Congressional science and technology fellows sometimes help. Constituents and interest groups of all stripes inundate Congress with fact, analysis, and opinion on every conceivable topic. Their inputs come via mail, e-mail, and telephone; through visits made to staff and members; in meetings held when members and staff return to home districts; and through a variety of other mechanisms, such as lunch and breakfast seminars. The biggest challenge is therefore not to find data, but to filter and analyze those data to create useful information.

Hearings provide the principal formal mechanism by which Congress gathers information. As Schneier and Gross (1993) note, hearings provide "a context that forces the confrontation of opposing viewpoints without doing violence to collegial relations. Through clever questioning of a witness, a legislator can challenge a colleague's perspectives without challenging the person. Meanwhile, he or she can learn what he needs to know through interpersonal communications, a type of exchange particularly congenial to the personality type of the politician."

Many technical experts who have been frequent participants in the hearing process have expressed a rather different view to us. When matters of complex science and technology are under consideration, these experts express doubt that the hearing mechanism reliably allows members and staff to develop a balanced and informed view of important technical issues. Indeed, with some complex technical issues, we know from personal experience that members

Box 1-1
Scientific and Technical Topics on Which Congress Needs Good Analysis for Informed Decisionmaking

- Choices about the focus and funding of federal programs to conduct or support research and development ...

  e.g., having doubled the NIH budget are there other changes needed in the balance among federal R&D investments? Are there effective ways to measure the productivity of federally funded research? What are the strengths and limitations of alternative strategies to encourage private sector research investments?

  ...and assess the benefits and other consequences likely to derive from such investments.

  e.g., will the tests of the planned antiballistic missile defense be sufficient to demonstrate its functionality and effectiveness? What kinds of scientific contributions can we expect from the International Space Station? How well can new oil extraction technologies and methods preserve wilderness environments in which they operate?

- Judgements about how available science and technology can best be brought to bear in addressing important social problems ...

  e.g., how cost-effective might biometric technologies be in securing our borders against unauthorized entry? How far can renewable energy technologies go in replacing conventional

and their legislative assistants find it difficult to even generate good questions for witnesses—and no questioner wants to risk looking foolish on C-SPAN. When Congress needs insight about longer term, larger scale problems, studies done by think tanks, university research groups, and the National Research Council (NRC) may be available. However, serious analytical coverage of the topics with which Congress must deal is often uneven. As Kingdon (1973) notes, "the voting decision for the congressman has been often portrayed as one in which there is either very little information available to the congressman upon which to base his decision or a plethora of information which is undistilled and hence useless."

Even when relevant expert input is readily available, it is not always welcome. Awkward facts can get in the way of compelling political considerations or can complicate the implementation of strongly held ideological agendas. The Kennedy School's Matthew Bunn was recently quoted as observing that "unbiased information does not serve ideologues well, either on the right or the left" (Mitchell 2002).

energy technologies? What technologies are available to assist persons with disabilities and integrate them into the economy and the community? How can the results of research on learning be best incorporated into the curriculum?

... or in providing insight that can help resolve contentious policy issues.

e.g., what is the role of anthropogenic sources in global climate change? How much arsenic is safe in drinking water? How important a determinant of human criminal behavior is genetics?

- Consideration of the potential societal implications of current scientific and technical developments and the regulatory and other needs that they may create ...

  e.g., developments in nanotechnology and biomaterials will augment human–machine integration. Should we care? What is our obligation to protect and preserve any life discovered on Mars or elsewhere beyond Earth? Will advanced computing and/or robotics cause technological unemployment?

  ... as well as emerging scientific and technical developments and the social responses they may require.

  e.g., what are the trade implications of new agricultural biotechnologies? What are the arguments for or against a moratorium or ban on human cloning? What are the national and economic security risks of reliance on the Internet?

Note: Developed with assistance from David H. Guston.

Information Is Not Knowledge

Information and facts are not the same thing as knowledge, understanding, and insight. Whereas some early commentators in science, technology, and public policy adopted a simple two-step model in which the technical community produces facts and the political community then uses them to make decisions (Kantrowitz 1975; Weinberg 1972), most serious contemporary students of science and technology policy analysis argue that when the problems are complex, raw facts are rarely of much use. An intervening step is needed, a step of balanced analysis and synthesis that sorts, integrates, and analyzes information to frame the issues and extract knowledge and insight. As Kingdon (1973) puts it, "Congressmen need not just information, but information that is usable. It must be predigested, explicitly evaluative information which takes into account the political as well as the policy implications of voting decisions." However, as Schneier and Gross (1993) note, "In the market for policy information there are few significant general suppliers or consumers.

The recipients of information have, so to speak, gone to bed with the suppliers." In short, careful, impartial, well-balanced analysis that is also sensitive to congressional needs is a scarce commodity. When it is available, demand may be limited.

Many problems that come before Congress cannot be adequately addressed without clear insights about the implications of key scientific and technical issues. Most of our technical colleagues share this view, as do most of the more technically oriented members and congressional staffers with whom we have held discussions. For example, what is the best way to manage the transition of telephone service from the current highly regulated conventional switched-line systems to the essentially unregulated packet-switched Internet? To answer that question, one needs to be able to estimate things like the effect of such a transition on the cross-subsidies from long-distance telephone calls that currently support rural telephone service. How should the benefits and risks of biotechnology (e.g., genetically modified organisms, therapies that use stem cells, and bioterrorism) best be managed to enhance human welfare and ensure the prosperity of U.S. industry? To answer such questions, one needs realistic estimates of the potential risks, costs, and benefits of such technologies and their associated uncertainties. What is the best way to evolve a reliable and effective air traffic control system for the coming decades? To answer that question, one needs to understand how well autonomous computer agents in aircraft engaged in free flight might be able to perform some of the functions now performed by ground-based human controllers. In short, it is unlikely that reasonable and balanced answers to these and hundreds of similar questions will be found unless the decisions are informed, at least in part, by the results of careful interdisciplinary analysis that has dealt in depth with the relevant technical details.

Developing such information requires much more time and expertise than most members or their staffs have available. Moreover, important issues involving science and technology typically cut across disciplines. Performing adequate analysis requires the integrated efforts of a number of different experts who have a variety of backgrounds and knowledge.

Limits to the Adversarial Process

Students of congressional decisionmaking tell us that Congress routinely makes decisions on subjects about which it has relatively little expertise. Indeed, from the perspective of members, the substantive details are often less important than broader political considerations. Schneier and Gross (1993) report, "Most of the important decisions that the Congress makes involve bargains and compromises and thus, in turn, strategies. It is not a system of policymaking for people who believe in revealed truths or in finding 'the one best way' to solve any given problem. Representative democracy is founded in a skepticism about such simple solutions and a faith in the utility of making deals."

In his study of congressional decisionmaking, Kingdon (1973) reports that when he asked "in good neutral interviewing fashion" how one congressman sorted out the conflicting scientific claims on a controversial issue, the member "snorted," "We don't. That's ridiculous. You have a general position. Once you assume that posture, you use the scientists' testimony as ammunition. The idea that a guy starts with a clean slate and weighs the evidence is absurd."

Whereas the inherently political nature of Congress makes such an approach inescapable, there are also inescapable physical and biological realities that govern the operation of the natural and social world. Fortunately, a legislature rarely directly sets out to contradict physical reality, as the Indiana State Legislature did in 1897, when, in a vote of 67 to 0 the House passed Bill No. 246, simplifying the value of π to 3.2 (Adams undated).

However, as the world becomes more complex and more technical, it becomes easier for processes based on negotiated deals and adversarial proceedings to produce results that ignore or do not adequately address scientific and technical reality. This is particularly true for decisions involving large, tightly coupled systems. Dartmouth's John Kemeny (1980) succinctly illustrated this point in his report as the chair of the presidential commission to examine the Three Mile Island nuclear disaster. Suppose, he wrote, "... that Congress designed an airplane, with each committee designing one component of it and an eleventh hour conference committee deciding how the various pieces should be put together?" Kemeny asks, would any sane person fly in such an airplane?

Of course, Congress does not design airplanes (although decisions about weapon system acquisition, space shuttle development, and similar issues have sometimes gotten close). However, Congress often gets involved in specifying how complex socio-technical systems should be designed, developed, or operated. When the elements of such systems are tightly coupled, an incrementalist approach to design and problem solving can lead to serious inefficiencies or ineffective outcomes.

An incrementalist decisionmaking procedure makes more sense than an impartial systems analysis if one is trying to balance interests in developing a complex piece of social legislation in which the key issues involve social equity, not system performance and efficiency. In a classic article on "the science of muddling through," Charles Lindblom (1959) outlined the strengths of the incremental adaptive approach to policymaking, which characterizes much of Congress's work. Kai Lee (1993) has made similar arguments in the context of environmental management, but in this case, because the natural world is the system being considered, he also stresses the importance of analysis as one of the inputs to deciding how next to move in an incremental, adaptive process.

In the current congressional environment, if informed decisions are to be made by uninformed members, stakeholders must take the initiative to come forward to state their views and concerns. Congress can then seek appropriate compromises. This adversarial system has tremendous advantages in a legislative context. First, because stakeholder groups are allowed and expected to make their case to their representatives in Congress, it is less likely that the needs of any group will be systematically ignored than it might be in a traditional scientific approach, where a few recognized and supposedly unbiased experts are charged with finding "optimal" solutions. Second, the generalists who inhabit Congress cannot possibly dig up what everyone needs, so a system where stakeholders come forward with this information is efficient. Third, by making stakeholder reactions part of the process, legislators not only learn what stakeholders prefer, they also learn the strength of those preferences.

Unfortunately, for some issues, regardless of the final decision, this approach fails to adequately illuminate all relevant considerations. Many of those issues involve science or technology. For a stakeholder-based process to be effective, stakeholders must be able and motivated to make their own case. This can be a problem when one interest group is large and diffuse, so that no individual member is highly motivated to act. The problem is even more acute when legislation is being considered that would create a new stakeholder, as is often the case with novel technology. Incumbent firms with much to lose from the introduction of new technology may be expected to mount vigorous blocking campaigns, but often there are no vocal proponents for new technologies, no academic or other group that has done systematic analysis of its potential benefits, and few in the general public who have appreciated the benefits that the innovation might bring.

For example, in the face of concerns about civil infrastructure security in the wake of the September 11, 2001, terrorist attacks, there has been much talk about the importance of strengthening the reliability of the electric power system. One of the best ways to do this would be by encouraging distributed generation and small-scale microgrids, operating below the level of the traditional distribution system. However, because legacy commercial players in the power industry do not view such developments as advancing their interests, and because many state regulators view the changes that would be needed to allow the introduction of such technologies as threatening their traditional prerogatives, no strong advocates for such an approach have emerged in recent policy debates on infrastructure security.

Another class of problem that requires technically based analysis is the assessment of unintended consequences. Proponents of new technologies and systems often are slow to think about or assess follow-on effects and "externalities." Good technical policy analysis can often reveal such consequences. For example, requiring schools to use technology that blocks porn from computer screens used on the Internet sounds like a great idea until you realize that perfect filtering is impossible, with the result that most filters also block sites that tell teenagers about AIDS. Members might still decide that on balance blocking should be required, but they should be able to make this judgement with a full understanding of the trade-offs. Several years after advocates of electric vehicles in California and several Northeast states had obtained legislative mandates intended to force their widespread adoption, independent analysts finally showed that recycling the half ton of lead contained in the batteries of such vehicles could release more lead to the environment than would be released if the cars were fueled with leaded gasoline (Lave et al. 1995). Of course, for some new technologies, such as genetically modified crops, opponents emerge in abundance, and the problem is not identifying potential unintended consequences, but rather sorting out which of the dozens of those that have been proposed are actually scientifically credible.

Analysis can also help to quantify the pros and cons in an argument. No congressional representative would or should abrogate his or her responsibility to make social choices to the results of technical analysis, but technical policy analysis can help them to make those choices by quantifying the magnitude of effects, costs, and benefits. People say in the abstract that Alaskan wilderness is more or less important than securing a larger domestic oil supply. In the process of balancing conflicting objectives, though, it helps if the quantity of oil, the number of jobs created, the nature and extent of ecological disruptions, and similar factors can all be estimated and the way in which those estimates depend on different assumptions can be illuminated.

The adversarial approach also depends on the ability of members of Congress to assess the arguments made by stakeholder groups, which can be difficult in highly technical matters. For example, Congress has considered various proposals on electromagnetic spectrum management that could reshape several major industries, but given that few members have scientific training, few understand adequately what "spectrum" is. Decisions may be based on analogies used by lobbyists, i.e., members imagine that "spectrum" is like land or like air. Analogies are useful, but ultimately "spectrum" is like neither of these things. Indeed, no analogy could possibly work for long because our understanding of how different systems can and cannot share a portion of the spectrum changes with technical innovation.

Confusion is sometimes greatest when there is significant scientific uncertainty. For example, members of Congress are sometimes genuinely perplexed that the scientific "facts" regarding global climate change may differ from one year to the next. Those without scientific training may not adequately appreciate the extent to which uncertainty is inherent in many of the conclusions of scientists. In an adversarial system, both sides can use uncertainty to obscure rather than illuminate. If your objective is to win a political argument, searching out an expert who shares your view may serve your ends. However, as Schneier and Gross argued in our earlier quotation from their analysis (1993), if "the desire to make good public policy is a significant variable" in members' decisionmaking, the availability of impartial, reliable knowledge and insight also becomes important. Even if it is not always used or not used in the way that analysts would like, its existence should over time help to move toward Brown's objective of operating "the machinery of American society with some efficiency, justice, and improvement" (Brown 1996).

In summary, there are at least six settings in which careful impartial analysis could assist in improving congressional decisionmaking:

1. when it is important to understand that a decision involves a tightly coupled system in which the traditional incrementalist approach to decisionmaking could easily lead to serious inefficiencies or ineffective outcomes;
2. when there is reason to believe that the inputs from traditional stakeholders and outside analysts may not adequately cover all the aspects of a problem that must be considered to determine the public interest;
3. when it is important to identify and evaluate possible unintended consequences or to evaluate those hypothesized by others;
4. when it is helpful to quantify the pros and cons in a policy choice;
5. when members and their staffs need help in understanding and interpreting the inputs of stakeholders and outside analysts; and
6. when members and their staffs need help understanding and interpreting scientific uncertainty that is important to a pending decision.

The fact that legislative decisionmaking in a democracy is fundamentally about weighing and balancing interests does not mean that there is no role for objective data, careful analysis, weighing of evidence, or the other activities of "rational decision making" (Hastie and Dawes 2001). High-quality analytical inputs may never be the primary determinant of congressional decisionmaking, but, like the process of stipulation in a legal proceeding, they can at least set some boundaries to the debate, rule out some scientifically impossible or incorrect arguments, and help to frame political decisions in technically defensible ways. Some of the most useful products from the old U.S. Office of Technology Assessment (OTA) were those that were cited by advocates on both sides in floor debates on how to proceed on a matter of policy.

The analytical tradition in American governance has been much stronger in the executive branch than in Congress. For the first 150 years of the United States, this analytical tradition was largely qualitative. Drawing on the analytical traditions of such classical endeavors as rhetoric, history, and the law, thoughtful people inside and outside of government structured arguments to advance the objectives of efficient and effective governance.

Although qualitative methods continue to predominate, the past 60 years have seen a steady rise in the use of more quantitative analytical methods. Qualitative methods proved inadequate for such pressing decisions as how best to clear the North Atlantic of Nazi submarines. The new field of operations research provided insights that often saved lives and improved operational efficiency. The post–World War II era saw the further rapid development of these tools as well as such related methods as decision analysis, risk analysis, and benefit–cost analysis. Today, a wide variety of both quantitative and qualitative methods for policy analysis are taught in academic programs all over the country and are routinely used by many government agencies in support of the decisions they make.

In many cases, Congress has mandated that executive branch agencies use a rational analytical decisionmaking procedure by collecting evidence and performing analysis before reaching a conclusion. How these activities are performed is in turn governed by more general rules laid down by Congress through the Administrative Procedures Act and several companion laws designed to ensure an open, balanced, and impartial decision process. The courts routinely review the operation of the entire process.

The Advisory Tradition

A number of executive branch agencies have developed considerable analytical capabilities, along with the ability to commission more sophisticated analysis from outside groups. In addition, a strong tradition has grown of seeking outside technical and scientific advice. Indeed, standing advisory committees have grown so ubiquitous and influential that Jasanoff (1990) has referred to them as "the fifth branch" of government.[1] Most of these advisory committees are authorized by legislation. In some cases, such as with the Science Advisory Board of the U.S. Environmental Protection Agency (EPA), that legislation also mentions providing advisory support to Congress, but typically such boards have not found a way to serve two masters and have remained creatures of the agencies they serve. With both internal and external sources of technical expertise, the executive branch has developed vastly greater analytical capabilities and resources than those available to Congress.

With military conflict becoming ever more dependent on science and technology, major wars have served as the stimulus for some of the more important changes in how government obtains advice on issues of science and technology. The National Academy of Sciences was established by hasty congressional action in the closing hours of the final congressional session before the Civil War. The National Academy of Sciences provided important advice during that conflict (Dupree 1957). The First World War saw the rise of the National Research Council (NRC) as an important advisory group. Whereas the NRC has mainly served the needs of the executive branch, in recent years Congress has often also turned to it for advice and assistance.

In addition to the NRC, World War I also saw the establishment of a variety of expert advisory committees (Dupree 1957). However, it was the Second World War that transformed the relationships among government, science, and technology (Buderi 1996; Bush 1970; Hart 1998; Kevles 1978; Sapolsky 1990; Zachary 1997). Formal sustained science and technology advice to the president began in the early 1940s, largely coordinated by Vannevar Bush. Over the subsequent decades, the arrangements became more and less formal and rose and fell in intensity, with the preferences of the current president and the needs of the time (Golden 1988; Hart 1998; Herken 1992; Smith 1992).

In the post–World War II period, the rapid proliferation of health, safety, and environmental regulation brought a need for executive branch regulatory agencies to obtain detailed scientific and technical advice to inform the process of risk assessment and management. Jasanoff (1990) has examined the role of such science advisory committees. Her analysis focused particularly on the EPA and the U.S. Food and Drug Administration. Similarly, Bruce L.R. Smith (1992) has reviewed the science advisory activities of the two largest and best established executive branch advisory committees, the EPA's Science Advisory Board and the Defense Science Board in the Department of Defense, as well as advisory committees at the Department of Energy, the National Aeronautics and Space Administration, and the Department of State.

Smith argues that executive branch scientific advisory panels can be classified into four categories:

1. peer review panels, whose main functions are to review research proposals, award fellowships, or make recommendations on individual grants;
2. program advisory committees, which are charged with strategic oversight of a scientific field or an agency's research and development program;
3. ad hoc fact-finding or investigating committees, which range from the more specifically technical to those investigating broader problems with important technical dimensions; and
4. standing committees that provide advice to decisionmakers on broad political and technical issues. This type may be found at the agency level, the bureau or subagency level, or the presidential level. (Smith 1992)

Smith notes that the third and fourth of these categories "... are more deeply and directly involved in broad policy matters.... These committees deal with policy issues that have important scientific or technical dimensions" (Smith 1992) or what Harvey Brooks (1964, 1988) has termed "science in policy" as opposed to "policy for science." Similar science-for-policy support is far less available to Congress.

Both Jasanoff (1990) and Smith (1992) conclude that the idea that standing groups provide science advice that is cleanly separated from policy and the broader political debate is clearly incorrect. Jasanoff (1990) describes successful executive branch science advisory entities as engaged in "boundary work" that involves a process of negotiation between the science and the broader agency and external political contexts. Smith writes the following:

> Science advisors who are effective ... almost always operate in the pragmatic rationalist mode. They bring to the advisory task or soon acquire a subtle understanding of how their efforts fit into the work of the agency or decision-maker they seek to advise. They may choose to write a report using the rhetoric of the utopian rationalist, but they will almost always have subtly negotiated the terms of what they will say so as to mesh with the goals of their clients ... (Smith 1992)

Whereas advisory systems for the executive office of the president and for executive branch agencies may prove most successful when they subtly accommodate the policy preferences of their clients, in the inherently bipartisan environment of Congress, such accommodation is far less practical. Indeed, given the periodic shifts in power between the two parties, no congressional support agency could expect to last for long if it regularly accommodated any particular political perspective. Nevertheless, agencies supporting

Congress must be just as attuned to the unique needs of their audiences as their executive branch counterparts if they are to deliver information in a useful form and in a timely manner.

Good Policy Analysis

In the preceding discussion, we made frequent mention of policy analysis. This raises the obvious question: what is "good" policy analysis? More than 20 years ago, in an editorial in *Science*, Morgan offered the following operational definition:

> Good policy analysis recognizes that physical truth may be poorly or incompletely known. Its objective is to evaluate, order, and structure incomplete knowledge so as to allow decisions to be made with as complete an understanding as possible of the current state of knowledge, its limitations and its implications. Like good science, good policy analysis does not draw hard conclusions unless they are warranted by unambiguous data or well-founded theoretical insight. Unlike good science, good policy analysis must deal with opinions, preferences and values, but it does so in ways that are open and explicit, and allow different people, with different opinions and values, to use the same analysis as an aid in making their own decisions. (Morgan 1978)

However, if governance in a democracy is about weighing and balancing interests, is it realistic to expect balanced and disinterested analysis to play a significant role? Won't people try to use analysis as a vehicle to support their arguments and as a club with which to beat on their opponents?

Of course they will, and they do. However, in a discussion of the various motivations that people and organizations have in undertaking policy-focused research and analysis, Morgan and Henrion (1990) argue that to be effective in political debate and in support of specific policy decisions, analysis must meet some minimal standards of being balanced and disinterested, that is, of being "substance focused." As the policy community has become increasingly sophisticated about the use and abuse of analysis and more attuned to the attributes of "good" analysis,[2] it has become progressively harder to use analysis in a narrowly adversarial manner.

Good policy analysis is only useful if the results are communicated effectively to policymakers. Interacting with policymakers is a specialized skill. Good scientists and engineers are trained to present technical information precisely and completely to each other. Those who try to communicate with Congress in the same manner are likely to encounter difficulties because the audience is fundamentally different. First, because Congress must address every policy issue that U.S. citizens can imagine, it is mostly populated with people whose knowledge is broad but not deep. Given the volume of material passing through Congress, members and staff have little time for unnecessary tutorials. Communications must clearly convey the background information needed by lay people for the issue at hand—no more and no less. Second, because Congress is an inherently political environment, members and staff need to understand details of how actions affect every stakeholder group, and they may be interested in legislative history. Congress must also focus on the achievable rather than the desirable. They need to understand how practical legislation and congressional oversight can affect an issue, with little regard to aspects of the problem that are beyond congressional authority, that require the deliberative body to act faster than is procedurally possible, or that require no activity at all within the current legislative calendar. Thus, communicating with Congress requires knowledge of Congress.

Seeking and Interpreting Science and Technology Advice

Whoever performs it, analysis and synthesis on policy problems in science and technology is valuable to Congress only if members, committees, and their staffs know how to ask for it and use it. This requires staff who have some basic technical knowledge—knowledge often not possessed by the lawyers and political scientists who make up the majority of members and Hill staffers.

However, some members and staff, through their formal educational backgrounds or through long experience, have sufficient expertise to know what questions to ask and how to ask them. Although they are still a small minority, their numbers have been gradually growing. One of the programs that has contributed significantly to this process is the science and engineering fellowship program, now supported by most of the major science and engineering professional societies and coordinated by the American Association for the Advancement of Science (AAAS). Since the early 1970s (Stine 1994), this program has supported approximately 1,300 Ph.D. level professionals, many of whom have served with Congress, either on members' office staffs or with committee staffs. Today, approximately 80 fellows serve per year, of whom approximately 30 serve in Congress, with the balance placed in various executive branch agencies. Many former fellows have stayed on after their terms ended and are now in permanent staff positions in Congress.

The fellowship program has drawn limited public attention, but fellows serving in congressional offices play an important role in helping members and committees identify the scientific and technical information they need and in searching for, evaluating, translating, and synthesizing that information into usable forms.

Does Congress Need New Institutions To Provide Science and Technology Advice?

Chapters 2, 3, and 4 discuss a number of past efforts by Congress to develop and use institutions that could provide it with serious balanced analysis on issues of science and technology. Central to much of this discussion is the U.S. Office of Technology Assessment (OTA), which was created in 1972 and operated until 1995, when it was defunded as part of a sweeping set of budget cuts that affected both the executive and legislative branches.

There is considerable disagreement both about how successful OTA was and why it was defunded. The arguments are explored at some length in subsequent chapters. The question for us here is does Congress need new institutions to provide it with science and technology advice, and if so, what kind of advice? Some commentators argue that there is no need, that current sources ranging from the Congressional Research Service (CRS) to constituents and stakeholder groups, and the analytical contributions of universities, think tanks such as Resources for the Future, American Enterprise Institute, and the Brookings Institution, and studies by the NRC adequately meet the need. In contrast, several members, as well as current and retired senior staffers from both the House and Senate with whom we recently discussed this issue, have a different view. They argue that CRS and other sources adequately serve most of the immediate needs Congress has for technical information and some of the needs it has for more detailed explanations. They further argue that the NRC, together with various think tanks and university analysis groups, meet some of the need for longer term, more comprehensive studies. However, they say, there remains a critical unmet need for midterm studies undertaken on time scales of 6 to 18 months, as well as more in-depth studies of critical issues with which Congress must deal that are not being adequately addressed by outside sources.

Whether Congress chooses to create additional institutions to provide it with balanced impartial advice on issues involving science and technology will be decided in the same way that Congress decides all issues—by weighing input from constituencies and various interest groups, talking with colleagues, and then balancing a range of substantive and political considerations.

The authors of this book believe unanimously that Congress and the nation would be better served through the creation of one or more new institutions designed to perform balanced nonpartisan analysis and synthesis for Congress on topics involving complex issues of science and technology. The balance of this book has two simple objectives:

1. to encourage and advance a broad discussion of this proposition, within Congress, among science and technology experts and their professional societies, by industry, and in the general public; and
2. to lay out and evaluate a range of alternative institutional arrangements that might be adopted—alone or in various combinations—to provide improved analytical advice to Congress on matters involving science and technology.

The four Chapters in Part II provide background for that discussion. In Chapter 2, Bruce L.R. Smith of Columbia University and Jeffrey K. Stine of the National Museum of American History of the Smithsonian Institution offer a broad historical and institutional perspective on how Congress has obtained and used science and technology advice in the past. Chapter 3, by Robert M. Margolis of the National Renewable Energy Laboratory and David H. Guston of Rutgers, the State University of New Jersey, provides a historical account of the OTA experience. Chapter 4, written by Guston alone, provides a more theoretical analysis of the OTA experience that draws insights from other contemporary developments.

While it operated, the OTA was widely admired by other legislative bodies around the world and served as the principal inspiration for the creation of similar advisory institutions in support of 13 European nations and the Council of Europe. In Chapter 5, Norman J. Vig of Carleton College reviews and draws insights from some of the European experience.

Part III of the book is devoted to developing and critiquing a range of alternative institutional models that might be used to provide improved science and technology advice to Congress. These models are not offered as mutually exclusive alternatives. Indeed, parts of several might together provide a better solution than any one taken alone. In Chapter 6, we explain the motivation for developing these models. Chapter 7, by Christopher T. Hill of George Mason University, explores the possibility of expanding analytical capability in one or more of the existing legislative branch agencies: the Congressional Research Service, the General Accounting Office (GAO), or the Congressional Budget Office. [A recent congressionally mandated experiment (GAO 2002) to see how well the GAO could perform assessments is evaluated in Appendix 3.] In Chapter 8, John Ahearne of Sigma Xi and Peter Blair of the National Research Council explore the issues that would be involved if Congress were to expand use of the National Academies complex. In Chapter 9, Albert H. Teich and Stephen J. Lita of the AAAS explore the possibility of enhancing the role of the highly successful program of science and technology congressional fellows. Chapter 10, which we have co-authored with MIT's Daniel E. Hastings, examines the feasibility of establishing a small organization within Congress that would farm out studies to a variety of groups around the country to perform analysis. Chapter 11, by Gerald L. Epstein of the Institute for Defense Analyses and Ashton B. Carter of Harvard's Kennedy School, examines the pros and cons of reestablishing a group to do analysis entirely within Congress. In Chapter 12, Caroline S. Wagner of the RAND Corporation and William A. Stiles Jr. of Old Dominion University explore the possibility of establishing a dedicated organization outside Congress, in much the way that RAND provides dedicated analytical support to the White House Office of Science and Technology Policy.

The final chapter, in Part IV of the book, asks "where do we go from here?" In it, we attempt to draw a series of insights from the discussions that have come before, make some general conclusions, and suggest a range of strategies that could move the country forward toward the goal of developing better institutional mechanisms to provide balanced nonpartisan science and technology advice to Congress.

Notes

[1]The first three branches are the traditional executive, legislative, and judicial branches, and the fourth is the independent regulatory agencies such as the Federal Trade Commission, the Securities Exchange Commission, and the Federal Energy Regulatory Commission. There is not unanimity on this terminology; other authors have nominated other groups, such as the press, as a fourth or fifth branch.

[2]In explicit recognition of the fact that the definition of "good" analysis is inherently normative, Morgan and Henrion (1990) offer the following "ten commandments" for good policy analysis:

1. Do your homework with literature, experts, and users.
2. Let the problem drive the analysis.
3. Make the analysis as simple as possible, but no simpler.
4. Identify all significant assumptions.
5. Be explicit about decision criteria and policy strategies.
6. Be explicit about uncertainties.
7. Perform systematic sensitivity and uncertainty analysis.
8. Iteratively refine the problem statement and the analysis.
9. Document clearly and completely.
10. Expose the work to peer review.

References

Adams, Cecil. (undated) *The Straight Dope.* See www.straightdope.com/classics/a3_341.html (accessed May 3, 2003). For additional details, see also www.urbanlegends.com/legal/pi_indiana.html (accessed May 3, 2003).

Brooks, Harvey. 1964. The Science Advisor. In *Scientists and National Policy-Making,* edited by Robert Gilpin and Christopher Wright. New York: Columbia University Press, 73–96.

——. 1988. Issues in High-Level Science Advising. In *Science and Technology Advice to the President, Congress, and Judiciary,* edited by William T. Golden. New York: Pergamon Press, 51–64.

Brown, Richard D. 1996. *The Strength of a People: The Idea of an Informed Citizenry in America, 1650–1870.* Chapel Hill: University of North Carolina Press.

Buderi, Robert. 1996. *The Invention That Changed the World: How a Small Group of Radar Pioneers Won the Second World War and Launched a Technical Revolution.* New York: Simon and Schuster.

Bush, Vannevar. 1970. *Pieces of the Action.* New York: William Morrow Company.

Dupree, A. Hunter. 1957. *Science in the Federal Government: A History of Politics and Activities.* Baltimore: Johns Hopkins University Press.

General Accounting Office (GAO). 2002. *Technology Assessment: Using Biometrics for Border Security.* GAO-03-174. Washington, DC: General Accounting Office.

Golden, William T. (ed.). 1988. *Science and Technology Advice to the President, Congress and Judiciary.* New York: Pergamon.

Hart, David M. 1998. *Forged Consensus: Science, Technology and Economic Policy in the United States, 1921–1953.* Princeton, NJ: Princeton University Press.

Hastie, Reid, and Robyn M. Dawes. 2001. *Rational Choice in an Uncertain World: The Psychology of Judgment and Decision Making*. Thousand Oaks, CA: Sage Publications.

Herken, Gregg. 1992. *Cardinal Choices: Presidential Science Advising from the Atomic Bomb to SDI*. New York: Oxford University Press.

Jasanoff, Sheila. 1990. *The Fifth Branch: Science Advisors as Policymakers*. Cambridge, MA: Harvard University Press.

Joint Committee on the Organization of Congress. 1993. *Organization of the Congress, U.S. Congress, December 1993*. www.house.gov/rules/jcoc2.htm (accessed May 3, 2003).

Kantrowitz, Arthur. 1975. Controlling Technology Democratically. *American Scientist* 63(September–October): 505–509.

Kemeny, John G. 1980. Saving American Democracy: The Lessons of Three Mile Island. *Technology Review* 83(June–July): 64–75.

Kenyon, Cecilia M. 1966. *The Antifederalists*. Indianapolis, IN: Bobbs-Merrill Company.

Kevles, Daniel J. 1978. *The Physicists: A History of a Scientific Community in Modern America*. New York: A.A. Knopf.

Kingdon, John W. 1973. *Congressmen's Voting Decisions*. New York: Harper and Row.

Lave, Lester B., Chris T. Hendrickson, and Francis C. McMichael. 1995. Environmental Impacts of Electric Automobiles. *Science* 268(May 19): 992–995.

Lee, Kai. 1993. *Compass and Gyroscope: Integrating Science and Politics for the Environment*. Washington, DC: Island Press.

Lindblom, Charles E. 1959. The Science of Muddling Through. *Public Administration Review* XIX(Winter): 517–529.

Mitchell, Dan. 2002. Congress Reassesses Tech Office. *Wired News*, August 7.

Morgan, M. Granger. 1978. Bad Science and Good Policy Analysis. *Science* 201(September 15): 971.

Morgan, M. Granger, and Max Henrion, with Mitchell Small. 1990. *Uncertainty: A Guide to Dealing with Uncertainty in Quantitative Risk and Policy Analysis*. New York: Cambridge University Press. (Paperback edition 1992. Latest printing (with revised Chapter 10) 1998.)

Sapolsky, Harvey M. 1990. *Science and the Navy: A History of the Office of Naval Research*. Princeton, NJ: Princeton University Press.

Schneier, Edward V., and Bertram Gross. 1993. *Legislative Strategy: Shaping Public Policy*. New York: St. Martin's Press.

Smith, Bruce L.R. 1992. *The Advisors: Scientists in the Policy Process*. Washington, DC: Brookings Institution.

Stine, Jeffrey K. 1994. *Twenty Years of Science in the Public Interest: A History of the Congressional Science and Engineering Fellowship Program*. Washington, DC: American Association for the Advancement of Science.

Weinberg, Alvin. 1972. Science and Trans-Science. *Minerva* 10(April): 209–222.

Zachary, G. Pascal. 1997. *Endless Frontier: Vannevar Bush, Engineer of the American Century*. New York: The Free Press.

Part II

Background

2

Technical Advice for Congress: Past Trends and Present Obstacles

Bruce L.R. Smith and Jeffrey K. Stine

Does the U.S. Congress need a new institution, set of institutions, procedures, kinds of expertise or access to experts? Should Congress in particular enhance its capacities to acquire and/or make effective use of scientific and technical advice? Congress, as a set of institutions, practices, and people, is mainly engaged in the task of acquiring and processing information from its environment. There is little that Congress does that is not aimed at acquiring information, processing it through its own unique prism of values, norms, and political judgements, and thus "adding value" to the inputs to produce the laws, resolutions, confirmations, symbolic behaviors, actions, and nonactions that constitute the outputs of the legislative process. Members of Congress rely for advice, first, on their own (invariably broad) circle of friends, acquaintances, and constituents, then on their colleagues, whose opinions they respect to greater or lesser degrees, and then on their staffs both individual and committee, on the Congress-wide support agencies, on experts from the executive branch with whom they and their staffs regularly interact, on outside experts whose views are solicited in hearings or informally, on pundits from the specialized and opinion media, and on the phalanxes of interest group and nongovernmental organization representatives and lobbyists who ceaselessly bombard them with information at every turn. It would be hard to argue credibly that in this decibel-happy information age the modern U.S. legislator lacks "advice" on any topic, scientific or otherwise.

The question must be posed in a nuanced fashion: is there a need for particular kinds of advice on particular issues in the light of present trends and institutional realities and a reasonable expectation that something useful can be done? Those of us who see a need to improve the use of expertise do not argue therefore for the existence of simple "gaps" that need to be "filled," like air rushing into a vacuum or water into a dry riverbed. The need is subtler—indeed, we must demonstrate that there *is* a need. We must also demonstrate that scientists and engineers can participate in some practical fashion in an increasingly fast and furious, ideologically charged, and clamorous congressional policy process in which the daily headlines demand almost instant responses from politicians. The requirement for science and technology advice is for "knowledge" and not mere "information." The desired product or process is analysis, synthesis, or, simply, understanding. Particularly important is the ability to reconcile partial perspectives into a useful whole. It goes without saying that wisdom—the combination of knowledge with practical understanding—is usually, if not always, in short or inadequate supply. (That has been so, at any rate, with all of the institutions we've been associated with!) The case for reform of how technical advice is provided to Congress must rest on the supposition that we stand a reasonable chance of producing the desired but difficult-to-achieve results through the proposed reforms and that the advice will be different from the stream of advocacy information that regularly inundates Congress. We must also adopt something equivalent to the medical maxim "first, do no harm," so that we do not merely add unnecessary layers to an already complex legislative process.[1] Certainly we must not create a nontransparent mechanism that could subvert democratic values. Good advice depends on the personal qualities and characteristics of individuals, and individuals who function well in an advisory role may be found in many different institutional settings.[2]

Because the goal is knowledge and not mere information, a lot of advice will fall short of the mark and can be redundant, a rehashing of the conventional wisdom, and unhelpful. When advice truly brings a new perspective or approach to a difficult problem, however, it is not only valuable but can be invaluable—priceless instead of worthless. The metric (or, if that term is too mechanical, the process) for evaluating advisory mechanisms is itself a most difficult and challenging task for the adviser and the advisee. Ultimately, the test of whether an advisory system works depends on those who are to be advised. Will they seek and will they pay any attention to the advice? Will they find it at all helpful within the practical constraints imposed by the legislative environment?

This book suggests that a public airing of what Congress deems to be the need for science and technology advice, and a thorough review of the myriad ways in which such advice can be provided, are the essential first steps toward any reform.[3] A vigorous debate of potential pitfalls and obstacles of any proposed reform, as well as a clear demonstration of the needs we seek to address, should precede any recommendations to the congressional leadership. A stealth advisory system for Congress is an oxymoron. Moreover, Congress has shown that it is quite capable of addressing the question of what kind of advisory assistance it may need; it has created, abolished, and modified institutional arrangements for staff support in the past. This book may be viewed as revisiting the themes of *Technical Information for Congress*, a report to Congress that was extensively debated a generation ago and that has precipitated a continuing discussion.[4] There are no settled truths, but it is useful to debate the issues in the light of past experience, as well as in terms of new and emerging needs.

This chapter does not discuss the virtues, advantages, or defects of any particular institutional arrangement; other chapters will explore the different models for providing advice and the practical problems that arise with each model. Our aim, rather, is to explore some of the general problems of providing scientific and technical advice to Congress, to sketch how the legislative branch has responded to its needs for special advice in the past, and, based on our look at the past, to suggest a few cautionary lessons to be kept in mind with any proposed new arrangements.

Scientific Advice in the Congressional Context

The scientific advice that Congress needs is in principle no different from any other kind of specialized advice—that is, an expert in some subject knows something that the generalist does not know, whether the subject is taxation, archaeology, microbiology, constitutional law, economics, tumor virology, or astronomy. Any subject will admit of virtually limitless division into ever more specialized subfields. The nature of science is to advance by breaking difficult questions down into more manageable questions that can be answered at some point, or in principle, empirically (though what constitutes adequate verification of theory is an issue on which philosophers of science may disagree). Congress, on the other hand, is usually not interested in learning more and more about the specifics of a technical issue; members of Congress want to know enough so that they can integrate various bits of specialized information into a broader context. Members of Congress are not interested in a physics or chemistry lesson; they absorb specialized information to apply it in a broad, value-laden context. Members of Congress may wish to learn from the leading specialists in a field but reserve for themselves the task of synthesizing the various inputs into a framework for decision. More commonly, however, elected representatives will find it convenient to deal with specialized inputs that have already been to some degree screened, synthesized, and packaged into a more manageable form for them. The filtered analyses are more easily digested by the members. The nature of Congress's work—the reverse of scientific endeavor—is to want to assemble parts into a whole, to blur the sharp edges of issues in the hope of achieving consensus and fostering compromise. Members of Congress seek agreement, not truth.

The policy issues facing Congress will usually involve the perspectives of more than one set of experts. Choosing among a multiplicity of views will be important. Indeed, on the more complex policy issues with significant technical content, experts from the same scientific field will often disagree. There are matters involving science and engineering where consensus is eventually reached (as on other matters) and where the affected interest groups, congressional staffs, and executive branch agencies all agree. Cooperation and agreement in many areas mark the normal business of the legislative branch, but cooperation is rarely newsworthy. However, even if contention is less frequent than one might infer from the media coverage of Congress, the issues on which Congress needs advice are generally not those where the constituencies, committee staffs, executive agencies, and the members themselves are in full agreement. The disagreements, moreover, frequently do not arise from ignorance but from tenaciously held, conflicting views, and from opposing priorities and policy preferences.[5]

To consider an example, in the 106th Congress more than 50 bills were introduced to deal with Internet policy, including taxation, privacy, barring children's access to pornography, regulation, intellectual property rights, and many other issues (Smith et al. 2001). Members may or may not have understood the technical aspects of the Internet, but they generally understood what they wanted to achieve. What blocked action was not ignorance of the issues but disagreement over priorities. What was the most important goal: To expand Internet use? To protect the privacy of users? Or should the nation first address the right of state and local governments to tax Internet commerce so as to provide a "level playing field" between e-commerce and more traditional forms of business? Or should one shelter the Internet from all regulation for a time until it has become firmly established? Are we at such a point already? Numerous committees had jurisdiction over some aspect of Internet use, but none had full jurisdiction. Should there be a comprehensive policy toward the Internet, a series of limited and partial measures (which could possibly work at cross-purposes), or no policy at all for the time being other than to keep hands off?[6] Members can readily grasp some Internet policy issues with little or no knowledge of the technical details, while other policy components will require some familiarity with the underlying technical dimensions. A quick briefing can often supply the need. In certain other respects, coming up with a policy framework that is coherent, sensible, and practicable may well call for a much deeper technical understanding. However, technical knowledge alone will not guarantee agreement on policy goals, and members may disagree on what constitutes good technical advice.

Advice that is useful to Congress usually meets a number of criteria. First, it must be *relevant* to the policy issue at hand. Members of Congress are not scientists and must quickly move past the purely analytical issues to the broader normative context of policy. One does not need to know how a refrigerator works to use a refrigerator, and one does not need to know all the technical workings of the Internet to formulate policy toward the Internet. A brief explanation of the technical aspects of a problem will frequently suffice for the congressional staffer or member of Congress. On other occasions, the technical details may be an integral part of developing an adequate policy. The Internet policy context, for example, clearly contains complicated side issues. Hence, a clarification of the overlaps and links among the different aspects of the whole problem may demand a deep technical understanding before Congress can formulate sensible policy for any part. The technical advice, whether a brief explanation or a thorough analysis, must come in a form that can be digested by Congress. Congress is a decentralized institution and will tend to absorb information in bits and pieces through its different receptors (i.e., its individual members and numerous committees and subcommittees). Sometimes, however, a case can be made for centralizing some analytic functions, and recommendations for jurisdictional changes may be an eventual product of a policy review (always a difficult task in practice and frequently unwise).

Second, the scientific advice that is most needed at the moment is that which strives to be *disinterested*. This is not an easy concept for a member of Congress to grasp because most politicians have a suspicion—born of a certain healthy realism—that scientists are not altogether immune from disciplinary, institutional, or other forms of self-interest.[7] There are enough *interested* perspectives on the Internet, to return to the example set forth above, including those held by technical experts from cable companies and equipment suppliers and from children's rights and privacy advocacy groups. The scientist adviser will add little value if he or she is merely one more interested party. Our political and judicial systems, however, make effective use of adversarial processes in reaching decisions. So we must be careful not to claim too much for neutral experts who are removed one step from the political battle. Clearly, however, another interested perspective would probably not resolve the current Internet impasse. The most useful advice for Congress in helping it move toward policies for the Internet will likely be disinterested.

A final criterion is that the advice should be *credible*. This concept includes both the criteria of relevance and disinterestedness, but it adds something more: the notion that the unit, organization, or individuals making a recommendation have done their homework, have proceeded in a fair-minded and thorough manner, have been in touch with all of the parties, and have a background or *gravitas* that makes them worthy of being heard. The idea is partly in conflict with the earlier criteria because, if the decisionmakers have a comfort level with the advisers, it usually means that they know where the group is "coming from," that is, what predispositions the advisers have acquired in the process of becoming accepted as credible analysts on an issue. Thus the advisers must have enough weight, familiarity with the issues, and acceptance from the parties to a dispute, without having too much prior standing and too established a niche in the political and institutional landscape. The good advisory group has a hard time working its way into the process and a hard time staying fresh and independent from longstanding institutional and political interests. Congress wants its advisory apparatus to be connected enough to be taken seriously by the interest groups but not so connected as to be merely an ally of one camp or another, and it would not want the advisory mechanism to bias the policy debate or constrain Congress's own freedom of action. Finally, because advising is a two-sided process—giving and receiving credible information—our reform ideas should not be limited to the legislators' side. How the scientific community gives advice and interacts with Congress may be as fruitful an area of reform as how Congress should position itself to receive advice.

Congressional Staffing

The U.S. Congress differs from all other legislative bodies of the world in the extent and depth of its staff resources.[8] The question of congressional staff is significant because staffers are critical to how Congress operates. Adequate staff resources enable Congress to exercise its full constitutional powers to organize hearings, conduct investigations, carry out oversight activities, vote appropriations, draft laws, and otherwise function.[9] Staff members heavily influence how Congress receives and processes technical information. Congressional staff support operates at three levels: the individual member's own staff, committee staffs, and the congressional support agencies. The House member typically employs, on his or her personal staff, 15 to 18 full-time staff members, who are divided between Washington and the home district. The senator's personal staff varies, depending on the size of the state: a senator from a small state might employ 25 staff, and a large-state senator some 45, again divided between Washington and the home state. Aside from these broad numbers, almost everything else varies in how members use their personal staff resources. The House member and the senator can also have some discretion in having fewer, more highly paid staffers, or more staffers who are paid less. Each office will have a penumbra of interns, volunteers, and possibly a science fellow from one of the professional societies to supplement the core staff. Staff members who work out of the home state or district are almost exclusively engaged in constituency services. In Washington, the representative or the senator will have a small number of legislative aides covering areas of responsibility that include issues with heavy science and technology content. (Almost all issues have *some* technical component, but certain high-profile issues are heavily influenced by scientific and technological considerations [Committee on Science and Technology 1981]).

The legislative assistants will usually have a broad portfolio, and their perspective is that of the generalist. The legislative assistant's familiarity with the technical aspects of issues will depend in part on the member's committee assignments. Some staffers will develop some degree of expertise on given issues, and the senator or representative may rely heavily on those staff members. Individual staffers tend to be intelligent, ambitious, and energetic recent graduates with liberal arts backgrounds; often they are from the home state or district. Many have been interns earlier or campaign volunteers. Like their bosses, they are political birds of passage, and, in contrast to committee staff discussed below, most tend to have short tenures and move on regularly—to interest groups, to executive branch agencies, to private companies, to campaign consultant firms, to return to school, or, in some cases, to seek elective office themselves.[10] Even the staff aides designated as legislative assistants will typically spend time on constituent services, answering mail, phone calls, and other nonlegislative duties, compressing the time available for "issues" or legislation.

The second level is that of committee staff. Most committees have between 45 and 70 staff members; the numbers grew spectacularly in the 1970s and 1980s and declined somewhat after 1994. Committee staffs are organized along party lines, with a larger number of majority party staff members reporting to the chairman of the relevant committee or subcommittee and a smaller number of minority party staffers assigned to the ranking minority member of the full committee or subcommittee. Staff ratios, like the ratios of Democrats and Republicans on committees, are determined by the party holding the majority, after consultations (or, more rarely, negotiations) with colleagues in the minority.[11] A powerful chairman can mobilize substantial staff resources and control a large domain. John Dingell (Democrat, Michigan), chairman of the huge House Energy Committee, for example, controlled some 145 staff members and a sprawling empire in the 1980s. The committee's staff had increased along with the expansion of its jurisdiction, resulting from the enlargement of federal energy programs during the 1970s and from the "subcommittee bill of rights" of 1973–1974, which gave augmented staff resources to subcommittee chairmen.[12] The availability of these extensive staff resources, combined with Dingell's adroit use of the media and his long experience, gave him an exceptional influence for a time. Such large committee staffs are now rare. However, a senator from a large state who is in the majority party and who chairs a major committee can easily still have 95 staff positions at his or her disposal and consequently will function as something of a mini-chief executive officer. Senators from small states command fewer, but still substantial, staff resources. The staff resources give the senator or representative the foundation for carrying a whole panoply of activities, including peppering the executive agencies with requests for information, studies, and the like. One or two staff members on the congressional side can occupy the time of larger numbers of executive officials in responding to requests for information and inquiries. Committee staffs tend to be somewhat more specialized than their counterparts in individual offices, though they are also generalists and rarely have great depth of scientific expertise. Certain committees—notably, the appropriations, intelligence, and tax committees and the Joint Economic Committee—operate, in effect, with career staffs with stable, long-term tenure much like an executive branch agency. Partisanship, although never wholly absent, is muted on these committees.

Staffs from the authorizing committees are less apt to have a career orientation, long-term tenure, and a bipartisan outlook.[13] In recent years, the tone of politics on Capitol Hill has become more shrill and ideological, and committee operations have been inevitably affected by this trend. Science and technology issues reflect the ideological tugs and pulls that divide Congress. A peculiar feature among some authorizing committee staff members (and their bosses) is that they are "technology enthusiasts." That is, they pursue some program or area of technology—such as fusion power or Internet training for preschoolers—with a crusader's zeal. On the other side are the antitechnology sharpshooters ever alert to real or imagined ill effects and/or the moral evils of pursuing a particular line of research or of technological development, such as fetal tissue research or genetic engineering of food crops. The deployment of committee staff resources in these cases to aid "the cause" may contribute ideological fervor but otherwise does little to advance understanding on complex technical issues.

Nevertheless, committee chairmen (and most members) are comfortable in dealing with "their own experts" on committee staffs and frequently feel that their staff resources are inadequate to very heavy job demands. Though much of the advice they receive comes through a cacophony of interest group and advocacy voices, members of Congress are more likely to "hear" technical advice from their own staff than from outsiders. Indeed, they count on their staff to screen out noise from signal in technical as with other matters. To put the point differently, members of Congress are bombarded with so much information, under such tight deadlines, that they count on their staff to select what is worth listening to and responding to. One of the more useful innovations in providing technical assistance to members of Congress has been the science and engineering fellows program launched in 1973 by the American Association for the Advancement of Science (AAAS) and several dozen other professional and disciplinary societies. Under this program, the professional societies provide a stipend to support a scientist, usually a recent Ph.D. or postdoctorate researcher from one of the physical or life sciences, who works directly for a representative or senator or on a committee for one year (Stine 1994). The science fellow fits into the normal working relationships of Congress and frequently provides valuable input into the decision-making process. Some science fellows have stayed on and made their careers as professional staff members.[14]

Although it was not typical to employ scientists and engineers as congressional staff before the 1960s, they were retained whenever there was a perceived need. During the 1940s, for example, Senator Harley M. Kilgore (Democrat, West Virginia) used his position on the Committee on Military Affairs to press for the formation of a federal science agency. As chairman of the Subcommittee on War Mobilization, Kilgore held a series of hearings, including one in 1945 in which he introduced legislation to create a national science foundation. The organization of these hearings and the drafting of the legislation were shaped in large part by the physicist Herbert Schimmel, a subcommittee staff member (Subcommittee on War Mobilization 1945; Maddox 1979).

The Joint Committee on Atomic Energy (JCAE), created in August 1946, was another arrangement that made extensive use of scientifically and technically trained staff. The senators and representatives on this committee were required to absorb and understand a great deal of technical information, and the committee therefore employed a number of scientists and engineers to undertake the essential staff role of acquiring, filtering, and translating that information for the committee's members. These staffers all held security clearances and traveled frequently to the nation's weapons facilities and national laboratories to gather information and prepare for site visits by the committee members. While in Washington, they wrote briefing papers, organized hearings, and met frequently with committee members (Heller 1991). The JCAE, with its monopoly of technical knowledge and the secrecy under which it operated, became too powerful, in the opinion of critics inside Congress and the executive branch. The whole arrangement, according to critics, served to bias energy policy too heavily in the direction of reliance on nuclear power. Critics also attacked the Joint Committee and the Atomic Energy Commission for failing to exercise adequate oversight and regulation of the nuclear industry. Eventually the JCAE was abolished, and its functions were distributed to separate energy committees in the House and Senate, along with a broad reorganization of the executive branch, which saw the transformation of the Atomic Energy Commission into two separate agencies: the Energy Research and Development Administration and the Nuclear Regulatory Commission (Green and Rosenthal 1961; Dyke 1989).[15]

Before the 1960s, aside from the unique Joint Committee on Atomic Energy, Congress also had no formal advisory mechanisms devoted specifically to the provision of scientific and technical information. In response to the successful earth satellite *Sputnik* in 1957 and the formation of the National Aeronautics and Space Administration (NASA) in the following year, Congress established its first science and technology-oriented committees: the House Committee on Science and Astronautics and the Senate Committee on Aeronautical and Space Science. Both committees hired small professional staffs conversant in the legislative matters within the committees' purviews. The House committee quickly sought to expand its jurisdiction beyond aeronautical and space considerations. Its chairman, Representative Overton Brooks (Democrat, Louisiana), expressed the committee's desire "to see a closer tie develop between the Congress and the scientific community" (Hechler 1980, 53).[16] Toward that end, he announced the creation of a Panel on Science and Technology in January 1960. The initial panel consisted of 14 prominent scientists and engineers from across the country who met periodically to provide the committee with informal advice on basic and applied science and on associated public policy concerns.[17] Unlike formal hearings, which were open and on the record, the Panel on Science and Technology's meetings were held in private with the committee members and staff, and no stenographic notes were taken. These seminar-style advisory meetings were kept off the record in the belief that this would lower inhibitions and raise the candor of exchanges. The committee paid the travel expenses and a modest honorarium for the panel members. Various aerospace contractors funded informal luncheons and receptions, in which other invited members of Congress were encouraged to exchange ideas with the panelists and an array of special guests. In all, the Panel on Science and Technology met 13 times between 1960 and 1972, when it was disbanded. Although useful in its time, the panel's support within the committee gradually faded. Managerially, the staff time needed for the proper arrangement and follow-up for these panel meetings seemed excessive in light of the committee's growing responsibilities, and not all of the committee's members found the format productive.

Moreover, the closed nature of the meetings and the financial contributions of aerospace contractors engendered public criticism of the arrangement, and the panel was never revived (Daddario 1967, 117; Hechler 1980, 50–60).[18] Outside the science committees, as the federal government's support of research and development accelerated across the board during the height of the Cold War, the sentiment developed within Congress to enlarge its own bipartisan advisory system, independent of the executive branch agencies, but agreement on how specifically to implement the goal proved difficult to achieve.

Congressional Support Agencies

Congress has at its disposal a set of staff support agencies, including the Congressional Research Service (CRS) of the Library of Congress, the General Accounting Office (GAO), the Congressional Budget Office (CBO), and—from 1972 to 1995—the Office of Technology Assessment (OTA).[19] The CRS had its origins in 1914 as the Legislative Reference Service (LRS) and took its modern form under the Legislative Reorganization Act of 1946. In 1949, the Library of Congress created a Division of Science and Technology to deal with the S&T reference requests that had proliferated after World War II. As the library's S&T holdings continued to mushroom during the 1950s, Librarian of Congress L. Quincy Mumford asked Congress for "a substantial increase in the Division's staff and for staff in the Legislative Reference Service to deal with congressional inquiries in these fields" (Mumford 1958). In the 1950s and early 1960s, the LRS played a role in helping to computerize congressional offices and improve their access to modern information systems and databases.[20] The LRS as a whole was called on for more substantial assistance as well in this early period. Although it was barred from issuing formal recommendations to Congress, LRS's background studies could present analyses that were clearly meant to suggest a general course of action or potential legislative measures.

In 1963, the House considered a bill to create a Congressional Science Advisory Staff, and the Senate debated a bill to establish a Congressional Office of Science and Technology. Although these first steps to form a specialized science and technology advisory mechanism for Congress failed, they nevertheless set the stage for an expansion of the Legislative Reference Service's efforts in this area, which in 1964 reorganized and formed a new Science Policy Research Division. The new division built up a core staff of scientific and technical experts, who provided useful technical advice to a number of congressional committees and individual members of Congress.[21]

The Legislative Reorganization Act of 1970 changed the name of the Legislative Reference Service to the Congressional Research Service. It also introduced a new emphasis on policy research and analysis. This was a big shift for the agency, which had previously devoted most of its analytic energies to law indices and digests, monitoring issue areas, preparing background materials on demand, and providing individual congressional members with quick responses to requests for factual information (Stine 1986, 50–52).[22] After 1970, CRS's policy research and analysis contributions resulted in longer reports, more extensive dialog with committee staff, and a variety of seminars and briefings for members and staff. These expanded activities and responsibilities led to a sharp increase in CRS staff size, which rose from 334 in 1970 to 805 in 1978 and to 893 in 1984, its peak in total staff.[23]

In the late 1960s and the 1970s the influence of the LRS and CRS on the Hill in a number of science and technology policy areas reached perhaps its high point. One of the more colorful and forceful figures was senior specialist Franklin P. Huddle. A redoubtable personality and energetic source of ideas, Huddle was a forceful advocate for new initiatives, though he never stretched too far the rules that forbade CRS staff members from making recommendations. The CRS was influential in generating hearings on science advice for Congress that led to the committee print *Technical Information for Congress* (CRS 1971) and in providing information and analysis used by the congressional advocates of the Office of Technology Assessment in 1973.[24] In addition, Huddle was an enthusiast for a national materials policy, which he tirelessly promoted. He and his CRS colleagues' background work helped pave the way for the passage of Title V of the Department of State Authorization for Fiscal Year 1979, which gave the State Department coordinating powers for all foreign technical assistance programs of the U.S. government and mandated the creation of a planning office in the Bureau of Oceans, International Environmental and Scientific Affairs (OES), and a science advisory committee to advise the Secretary of State.[25]

In the 1980s the CRS continued to enjoy good standing and modest influence in science and technology matters, generally operating in a lower key fashion than was the case with Huddle, who from time to time ruffled some feathers on the Hill. More recently, the CRS role in science affairs has appeared to diminish, with the science policy staff being reorganized and disbanded as a separate unit. However, this may be more appearance than reality. For many of the component specialties that were previously included under the broad rubric of "science policy"—such as information technology, telecommunications, environmental quality, energy, bioethics, space, and others—have simply been dispersed into different organizational units and continue to be represented as the CRS repackages itself in response to the shifting tides of congressional interest and political fashion.[26] "Science policy" has always been a conglomerate term, a protean concept encompassing and generating numerous subthemes and issues. The whole spawns subparts, which evolve an intellectual life and a policy focus in their own right. The task of the parent discipline is to bring some kind of intellectual order and policy coherence to the overlapping and clashing offspring.

CRS as a whole continues to enjoy modest respect on the Hill, while the fortunes of the individual divisions wax and wane like those of academic departments. Whereas science policy seems to have disappeared from the agency's organizational chart, many of the major topics previously embraced by that heading continue to be analyzed as part of environment, energy, and health policy research. The American Law Division has retained its organizational identity and remains one of the strongest and most widely respected CRS units; it is regularly called on by committee and individual member staffs. The CRS has largely escaped the budget-cutting axe that has descended on other Congress-wide staff units. CRS's sustained congressional support probably stems from the fact that the agency has always provided direct support to individual members of Congress (who acquire information for their speeches, to assist in constituent services, and the like), not just to committees and subcommittees.[27]

The General Accounting Office, created by the Budget and Accounting Act of 1921, has provided advice to Congress on a variety of topics over the years.[28] In the 1930s, under powerful Comptroller General John Raymond McCarl, the GAO was a force to be reckoned with on executive branch operations, but the influence was of a negative sort: preaudits and audits of individual expenditure vouchers sometimes snarled executive operations in red tape. After World War II, Congress substantially recast the GAO's mission to focus on postaudit scrutiny and examination of executive agency accounting systems as a whole rather than on individual vouchers, a step that inevitably reduced substantially the size of the GAO's staff.[29]

Under Comptroller General Elmer B. Staats (1966–1981), the GAO sought a role beyond that of auditing and accounting narrowly defined to a broader program audit and policy analysis function. Staats was urged in that direction by Congress itself, which was looking for ways to strengthen its oversight capabilities. Toward that end, both the Legislative Reorganization Act of 1970 and the Congressional Budget and Impoundment Control Act of 1974 gave GAO more resources and latitude (Mosher 1979, 187; Wells 1992, 56). With this newfound backing, Staats created a program analysis division in 1976, headed by Harry G. Havens, which functioned as a staff unit and reported directly to the comptroller general. Staats was partially successful in transforming the GAO work product to include program audits as well as financial audits. But many in Congress were not entirely comfortable with a policy analysis role for the GAO, fearing a dilution in its traditional auditing capacities. As one senior member of Congress remarked to one of us, "The GAO couldn't find a marble in a teacup." The program analysis division was, in any event, reorganized out of existence as a separate unit in 1981, and its functions were consolidated within a larger line division. It continued to provide methodological and analytical assistance to other divisions.

Staats had a strong personal interest in science policy, stemming from his days as assistant director and then deputy director of the U.S. Bureau of the Budget. Yet he never developed a staff unit specifically labeled as a science and technology policy analysis office, preferring initially to rely on the CRS science policy resources staff for assistance when program audits required scientific expertise or to direct congressional requests directly to the CRS. Many program audits, however, in areas ranging from health to national security to environmental regulation, dealt with issues that would have significant scientific and technological components. In 1972, Staats identified S&T as an "issue area" to help address the need for technical inputs in GAO program audits, a move that allowed agency managers to assign up to 24 staff members for temporary periods to science policy studies of broad relevance to other ongoing GAO activities. Following Staats's departure in 1981, S&T was downgraded from an "issue area" to the category "area of interest," which reduced the staffing dedicated to S&T studies. In 1988, the area of science and technology was totally eliminated as a category within the agency's organizational chart.[30]

The GAO, as the largest of the central staff units serving Congress, has always been a potential target for criticism and has often come under close scrutiny from watchdogs on Capitol Hill. This has been a reality for any staff unit, especially one with a large budget and staff, which operates outside the direct control of members of Congress. The "natural" route for members to receive technical advice is through their own and their committee staffs, and the Congress-wide staff units spend much of their time briefing committee and members' staffs. Members chafe at what seem to them inadequate staff resources for their own offices and committees, while having to vote large appropriations for the GAO. The GAO traditionally enjoyed a fairly narrow base of congressional support—principally, the government affairs committees—and largely limited its work to responding to requests for audits and studies from committee chairmen rather than from individual members.

In the early 1990s, partly under congressional pressure and partly in a preemptive move to forestall even deeper cuts, the GAO launched a streamlining reorganization that initially reduced its work force from 4,958 in 1993 to 4,342 in 1995, before it was required to lower its staff further to 3,275 by 1999 (Ornstein et al. 2000, 129, table 5-1). In recent years, the character of GAO's work has appeared to shift somewhat toward the more traditional financial auditing function, though the agency has by no means abandoned the broader program audit. From time to time, the GAO provides advice on technical issues to congressional members and staff, and Congress in principle could direct the GAO to augment its technical capacities either by providing new resources or by reprogramming current efforts.

The Congressional Budget Office, established by the Congressional Budget and Impoundment Control Act of 1974, fills a specialized niche in the panoply of congressional support agencies. Its functions of cost estimation, "scorekeeping," and projections of budget deficits and surpluses are delineated and mandated steps in the budget reforms enacted by Congress and provide a stable source of funding and core mission for the agency. The CBO has earned a reputation for professional competence and nonpartisanship; it has remained small, generally having about 130 full-time equivalent positions. Unless Congress were to change its budget process or the agency's performance were to decline sharply, the CBO's future seems assured.[31]

The Office of Technology Assessment, which operated from 1972 to 1995 as a major source of technical advice to Congress, will be discussed at greater length in a later section. Suffice to note for the present that, when the incoming Republican majority abolished the OTA after its victory in the 1994 congressional elections, it did so in part for the same reasons that the GAO was trimmed. The OTA was regarded as a costly, unresponsive staff unit that worked on its own rather than on a congressional time schedule. In addition, the mood in Washington had become much more ideologically charged, with the Republican *Contract with America* shaping the congressional agenda and the struggle surrounding the Clinton health plan still fresh in people's minds (Gillespie and Schellhas 1994). Advice from advocacy groups representing ideological positions seemed to be the kind of advice that suited the times. The OTA was faulted by the new congressional majority for its slow pace, the irrelevance of many of its studies, and as a hotbed of industrial policy advocacy on behalf of Democrats in previous Congresses.

Congress and the executive branch, in the Clinton–Gingrich era of divided government, frequently clashed on technical issues, as on other issues. There was limited scope for neutral, nonpartisan technical advice from the scientific community in such circumstances. The scientific community has been undecided over, first, whether and how far it has a responsibility to encourage (and, if so, to prepare) its members to give technical advice to any part of the government and, second, whether to take stands on public issues apart from those of direct professional relevance to their members. Technical issues were hotly debated throughout the 1990s but often along partisan, ideological lines. Scientists could participate as individuals, lending their voices to causes in which they believed, but many scientists felt uncomfortable becoming part of the cacophony of advocacy groups pressing their causes to Congress. They no longer were functioning as scientists but as partisans. Scientists as citizens have every right to join in public debate, and they often have been encouraged to be more vocal (Committee on Science 1998). Some of the professional societies recently have begun to debate steps that might be taken to encourage and prepare some of their members to testify at hearings, hold special seminars and hearings, and otherwise interact with members and committee staff on technical issues.[32] The aim is to encourage the development of practices to involve scientists in the policy process in ways that make full use of their professionalism and objectivity.

The Role of the National Academy of Sciences

Congress has also turned to the National Academy of Sciences (NAS; now the National Academies complex) for advice on matters dealing with science and technology from time to time beginning in the 1960s (Stine 1988). Since its inception during the Civil War, the NAS had advised the federal government, but such work had been traditionally oriented toward the executive branch.[33] In 1961, the NAS established the Committee on Government Relations (renamed Committee on Science, Engineering and Public Policy [COSEPUP] in 1963) to expand its participation in the formation of national science policy and to provide advice on the status and needs of particular scientific disciplines. George B. Kistiakowsky, science adviser to President Eisenhower, was the committee's founding chairman; he cultivated close working relationships with several federal agencies. The NAS leadership was also interested in strengthening the organization's ties with Congress to enhance its mission of providing scientific advice to policymakers. A formal agreement was reached in December 1963 in which the NAS would advise Congress on science policy issues—specifically, the COSEPUP was to undertake a series of studies for the newly created House Subcommittee on Science, Research, and Development. The studies were to be financed by Congress and made available to all interested congressional committees (and ultimately published).[34]

Despite the usefulness of these studies and the links with the scientific community that the arrangement provided, the relationship between COSEPUP and Congress proved to be ephemeral. The failure to develop close and sustained working relations between Congress and NAS can be attributed to the fact that the NAS has historically provided advisory services primarily to executive agencies and to the practical difficulty that congressional committees have usually lacked the resources to pay for outside studies. This has been a fact of life for NAS operations. There also has been some feeling in Congress that the pace of NAS operations and the formality of its procedures did not always match well with the style of congressional operations. Congress has, however, continued to turn to the NAS indirectly, insofar as it frequently mandates executive branch agencies to contract with the National Academies for studies in environmental risk assessment, water pollution, auto safety, and numerous other technical areas. In recent years, Congress has typically mandated some 20–40 studies per Congress to be conducted by the NAS, a majority of the requests coming in the second session of Congress.[35] The role of the NAS has been considered most appropriate when a high-end, "Cadillac" treatment of an issue is called for and the prestige and depth of talent available to the Academies are thus particularly suitable. Congress has also acted to protect the advisory role of the NAS by amending the Federal Advisory Committee Act to provide it some relief from the potentially adverse consequences of lawsuits brought to open up the operations of the National Academies to wider public scrutiny and to impose on the NAS the open meetings requirements applicable to federal agencies.[36]

The Office of Technology Assessment

The experience of the Office of Technology Assessment, to begin with, can be conveniently addressed by posing three questions:

- What were Congress's original intentions in establishing the office in 1972?
- How did its day-to-day functioning relate to congressional activities?
- What were the causes of OTA's abolition in 1995?

From the outset the OTA suffered from a lack of clarity as to what its role should be. Congress in the 1960s debated the establishment of a Congressional Office of Science and Technology but came to no clear resolution of what need would be served and did not implement the idea. The enthusiasm behind the idea for the OTA in 1972 stemmed originally more from the congressional support agencies and from the outside scientific community than from Congress itself. Only a relatively narrow base of congressional support materialized. The early association of the OTA with Senator Edward Kennedy (Democrat, Massachusetts) created the impression, however unfairly, that political motives were behind the creation of the office. In actuality, the OTA mission was narrowly circumscribed in its charter (see Appendix 1): the agency was forbidden to issue recommendations to Congress but was instead directed to make presumably neutral background studies and analyses. The governing Technology Assessment Board of representatives and senators was evenly divided between the two political parties to help lessen the suspicions of partisan motivations. But what, exactly, was the OTA to study? No precise definition of an "assessment" was ever to emerge, but the words "technology" and "assessment" were wisely ignored in favor of a broad definition encompassing any significant issue involving scientific and technological developments with which Congress had to deal. Nonetheless, under its first director, Emilio Q. Daddario, a former Democratic congressman from Connecticut who had been a prominent member of the House Committee on Science and Technology and an early advocate for creating a congressional office of technology assessment, the OTA struggled to avoid being abolished in the face of budgetary stringency. Survival in the early years was an accomplishment attributable to Daddario's political skills. Internal quarreling among members of the staff over its mission and personality clashes also plagued the office in its formative years.[37]

Russell W. Peterson, a former Republican governor of Delaware who held a doctorate in physical chemistry and who had spent his early career working for DuPont, took the reins as OTA's second director. He sought to calm the political turmoil surrounding the agency by launching an elaborate and time-consuming effort to tap the broad scientific community for ideas on the OTA's work plan and appropriate goals. The effort did not result in a clear management or policy orientation for the agency, but Peterson's term did establish the principle of bipartisan leadership for the office.[38] Peterson also reorganized the staff to place program operations in the hands of technically competent directors, and he formalized recruitment practices to stress the hiring of individuals with training in the hard sciences. More than 50% of OTA's permanent staff eventually had doctoral degrees in science, engineering, or medicine, and political party affiliation played no part in hiring. Furthermore, Peterson instituted the practice of requiring the support of the ranking minority member of a committee before the OTA would undertake a study. (The OTA responded to requests for studies only from committee chairs, not from individual members, a practice that had the unintended consequence of limiting OTA's contacts with other members of Congress and their staffs.)

Under its third director, John Gibbons, a scientist from Tennessee with extensive national laboratory and science policy experience (and with service on the staff of the last energy office in the Nixon White House), the OTA evolved into a competent applied research organization with high standards, strong ties to the scientific community, a solid professional staff, and a reputation for conducting thorough studies. In some respects, the agency functioned like the NAS in drawing on outside scientists who served on a pro bono basis on committees backing up the work of the OTA staff. OTA studies and procedures, like those of the National Academies, could be highly structured, ponderous, and slow moving. However, the contrasts with the NAS were more notable. OTA volunteer committees were advisory only to the OTA staff who wrote the reports (in contrast, NAS's committees are the authors of Academy reports, the staff their instrument). OTA panels were selected to hear all viewpoints (NAS's panels try to achieve consensus). Reports of the NAS invariably issue recommendations, whereas OTA's reports rarely issued explicit recommendations. The OTA work product, nonetheless, was deemed by critics to be too voluminous (reports were frequently lengthy) and submitted too late to be truly responsive to congressional needs.[39]

The Republicans generally chafed under the long period of Democratic congressional hegemony and felt short-changed on staff requests and the allotment of committee seats by what they saw as an increasingly arrogant majority leadership. They became indignant in the 1980s over a series of OTA studies that seemed to them to advocate a highly partisan set of industrial policy goals and measures.[40] Policies reflecting these ideas were incorporated into the 1988 Trade Act, including the creation of the Advanced Technology Program (ATP) in the Department of Commerce, the renaming of the National Bureau of Standards as the National Institute of Standards and Technology (NIST), and enhanced trade protections. Republicans in Congress strenuously opposed such measures as protectionist in intent and unwise subsidies, or "corporate welfare" measures that pushed the government inappropriately into the marketplace. The Reagan–Bush administration reluctantly accepted the initiatives of the Trade Act as the price of doing business with a Congress controlled by the Democrats. For the OTA the result was a paradox: relevance was achieved by participating in the congressional policymaking process but at the cost of alienating many in the minority party. OTA studies of the Reagan "Star Wars" proposals were perhaps even more significant in shaping critical reactions toward the agency among Republicans.[41]

The criticisms were not always consistent: the OTA was assailed, on the one hand, for irrelevance and, on the other, for being too influential in a partisan direction. Nor were only Republicans irritated by OTA studies. OTA studies of deregulation of the electric power industry, automobile fuel economy standards, and environmental risk assessment angered some Democrats as much as the agency's assessments of Star Wars and ATP angered some Republicans. The OTA in the 104th Congress issued a study critical of its earlier industrial policy enthusiasms, but as the ideological controversies heated up, there was no solid base of support in Congress protecting the agency. The OTA enjoyed only limited backing in Congress, and this support principally came from a small circle in the majority Democratic party and an even smaller group of ranking minority Republican members.[42] The chief Republican supporters were, on the Senate side, Ted Stevens of Alaska (who chaired the Technology Assessment Board when the Republicans controlled the Senate during President Reagan's first term), Orrin Hatch of Utah, and Charles E. (Chuck) Grassley of Iowa and, on the House of Representatives side, Amo Houghton of New York.

When the Republicans swept into control of Congress in the 1994 midterm elections, it soon became evident that the OTA was in trouble. The Republicans had talked during the campaign, which prominently featured Newt Gingrich's *Contract with America*, of reducing congressional staffing, abolishing a number of federal agencies (such as the Department of Energy and the Department of Education), and sharply curtailing or eliminating various programs (including the Advanced Technology Program in the Commerce Department, the Technology Reinvestment Program in Defense, and legal aid). They had also promised not to spare their own congressional prerogatives from critical scrutiny. When they encountered resistance to the more ambitious efforts (or had second thoughts about the wisdom of some of the proposed measures), it was natural to look for a target that could be more readily achieved. The OTA was an ideal candidate: it was a weak opponent with thin support within Congress; its abolition would fulfill a pledge to reduce congressional staff; and no significant outside constituency group would be offended (the almost total lack of reaction to OTA's demise from the scientific community and the various professional societies appeared to be ample confirmation of this judgement[43]). In retrospect, it may be wondered why the agency lasted as long as it did. It was founded in controversy, struggled to find a mission and a secure niche in the congressional scheme of things (and only partially succeeded), had limited support within Congress even if it was admired by many in the outside scientific community and emulated by several foreign legislatures (Vig 2003; Vig and Paschen 2000), became a focal point of partisan dispute whether unfairly or not, and tried to function as a kind of internal "think tank" at a time when sharp ideological cross-currents were increasingly buffeting congressional operations.

Could or should an OTA be reestablished to provide timely and relevant technical advice to Congress? In strict terms, the OTA was not abolished since its authorizing statute was never specifically repealed; it was simply defunded in 1995 (and no appropriation has been voted since then, notwithstanding repeated efforts by Representative Rush Holt [Democrat, New Jersey] to appropriate funds to revive the office). A last-ditch compromise proposal by Representative Vic Fazio (Democrat, California) to salvage a pared-down OTA in the 104th Congress by placing it in the Congressional Research Service failed at the time but presumably could be revived if it had any support. Any effort to recreate the agency would have to overcome a built-in resistance to the formation of any new congressional bureaucracy on the part of most Republican and many Democratic members of Congress, and any new mechanism might also have to counter a presumption of past failure that may linger from the OTA's demise.[44]

Conclusion

Congress has evolved a flexible system for tapping outside technical advice that has reflected its changing needs and shifts in the congressional agenda and in the wider policy climate. It is also clear that Congress has drawn on its own committees, its staff support agencies, and on a variety of contacts with outside bodies, nongovernmental organizations, and the executive branch agencies to obtain the scientific advice it has deemed necessary, modifying its practices in the light of new circumstances. The institutional fortunes and the capacities of various advisory units have waxed and waned; a staff strong at one time became less so at another. Organizations have come and gone. The external climate in which Congress operates has also undergone dramatic changes that have inevitably affected how Congress reaches out for or receives advice. In all of this, Congress has clearly not lacked technical advice or access to such advice in any simple fashion, although it may have from time to time lacked the best advice or felt itself poorly served by its various advisers. There is always a need for ideas that are timely, disinterested, practicable, comprehensive, and wise, but, like virtue, they are not easily summoned.

The velocity of ideas, requests, proposals, demands, and recommendations is staggering, as is the workload of the modern Congress. One senses that there is something amiss with the system as it currently operates. What can be done to improve the way that Congress obtains scientific advice? One can begin with how the executive branch agencies operate and interact with their congressional counterparts. A less partisan atmosphere in Washington may invite experimentation in ways to improve technical communication with Congress and its constituent units. The outside scientific community, as represented in the professional and disciplinary societies, might consider a host of ways to alert members to the opportunities for interaction with Congress and also consider ways to prepare scientists for effective presentations of technical information to Congress. Scientific career paths might be modified to create incentives for some fraction of working scientists to devote serious efforts to the public (and congressional) understanding of science.[45] Significant steps could be taken without converting scientists into an army of partisans and political activists, a prospect that would be distasteful to most scientists as well as injurious to the objectivity, detachment, and quest for truth that characterize the scientific enterprise. Of course, advocacy-tinged science can play a useful role in clarifying policy choices for members of Congress, too, and individual scientists who wish to influence the policy process may choose to participate on behalf of a cause they believe in as citizens. The "think tank" policy community in Washington might augment its capacity to play a role in the debate of scientific issues by adding scientists and science policy specialists to their staffs.[46]

To help understand and improve how the government uses scientific advice in the policy process, the Carnegie Corporation of New York established the Carnegie Commission on Science, Technology, and Government in 1988. The commission was divided into committees that addressed all three branches of the federal government. The Committee on Science, Technology, and Congress published three reports that outlined various recommendations for how Congress and the scientific and engineering communities could improve their interactions (Committee on Science, Technology and Congress 1991a; 1991b; 1994). Many other reform proposals have been advanced recently, including the various models discussed in Chapters 7–12 of this book. We leave to our co-authors the discussion of the specific reforms, but we note that it is neither likely nor desirable to expect Congress to change how it operates to accommodate scientists or any other group. No one should want to create privileged access for any group to the inner councils of congressional policymaking or to dilute Congress's democratic accountability.

Congress can, however, improve its access to scientific advice without fundamentally changing how it operates, starting with modest steps on where hearings are held, additional meetings with scientific groups, and other measures. Congressional staffs can be augmented, and scientific expertise within existing support agencies can be strengthened.[47] An historical perspective should caution us against the search for silver bullets but should encourage experimentation based on evidence and argument. For tinkering with our institutions by applying practical reason is very much in the spirit of the new "science of politics" championed by the founding fathers.

Acknowledgements

The authors are indebted to the following people for the comments they provided on the first draft of this paper: Peter Blair, William Frenzel, David H. Guston, Genevieve J. Knezo, James Linsay, M. Granger Morgan, James H. Paul, Jon M. Peha, and James A. Thurber.

Notes

[1]William D. Carey once observed, "To help Congress deal effectively with science, it is not ... necessary to foist a spate of new committees or institutional innovations on an overloaded and overstructured legislative body." See his comments in Orlans (1968, 266). Also useful are Lakoff (1974); *Science and the Congress* (1978); and Abelson (1988).

[2]For a general discussion of personal qualities that promote good advice-giving and institutional arrangements that help or hinder the use of expertise, see Meltsner (1990) and Smith (1992, chap. 9).

[3]Our use of the term "science and technology advice" is generally meant to refer to the physical, life, and cognitive sciences and to the related engineering and biotechnology disciplines.

[4]The report (CRS 1971) consisted of 16 case studies of congressional decisionmaking involving science and technology (S&T) issues, or broad policy issues containing a heavy S&T component, with a brief discussion of the implications for effective use of science advice at the end of each chapter. The definition of S&T issues was broad,

arguably too broad; e.g., a chapter on the establishment of the Peace Corps might not appear to all observers as an illustrative use of technical advice.

[5]Whereas technical analysis can help frame problems and set limits on the possible, science cannot tell Congress what should or should not be done. Such choices involve values and are ultimately political decisions.

[6]Even the regulation of scientific research itself involves a similar constellation of variables. See OTA (1986).

[7]We recommend a reading of any congressional hearing on "scientific pork" or "peer review" to illustrate the point that "peers" do not mean quite the same thing to a scientist and to a member of Congress.

[8]Total congressional staffing reached a high point of 28,031 in 1987 and fell to 23,648 by 1999 (Ornstein et al. 2000, 129, table 5-1). Until its turning point in the late 1980s, the overall number of congressional staff had risen since the 1950s, largely in response to the steadily expanding workload for members of Congress, especially with regard to increases in the number of committee and subcommittee hearings, the amount of time spent in session, the number of pages of public bills enacted, the volume of constituent correspondence, and the demands of campaign fund-raising (Ornstein et al. 2000, chap. 6).

[9]Since 1970, Congress has significantly increased its oversight activities, largely in response to the continued growth in federal spending and to the expanded role of the presidency under Johnson and Nixon. The Legislative Reorganization Act of 1970 formalized congressional oversight by requiring most committees to prepare biennial reports on their oversight activities. The act also authorized the retention of additional committee staff, and it strengthened the ability of the Congressional Research Service and General Accounting Office to assist committees in their evaluations of executive branch agencies. Many committees responded to the act by creating oversight subcommittees (Ogul 1976; Foreman 1988; Aberbach 1990; LaFollette 1990).

[10]For general treatments of congressional staff, see Fox and Hammond (1977); Kofmehl (1977); Malbin (1980); Rundquist et al. (1992); and Hammond (1994). For staff retention, see Salisbury and Shepsle (1981) and Henschen and Sidlow (1986).

[11]The initial 50–50 split in the Senate in the 107th Congress gave the Democrats grounds for arguing for substantial parity in staff ratios between the parties on Senate committees.

[12]In addition to the full committee, Dingell chaired the Subcommittee on Oversight, which blazed a remarkably broad investigatory trail. For a discussion of the 1973–1974 changes in committee resource allocations, see Deering and Smith (1985). For a general treatment of staff resources, see Ornstein et al. (2000, chap. 5 and accompanying tables).

[13]During its formative years, the newly created House Committee on Science and Technology proved to be an exception to this general trend (Hechler 1980).

[14]The career of Robert Palmer is illustrative. A biologist who became a AAAS science fellow in 1979–1980, Palmer served on the staff of the House Science Committee and stayed on after his fellowship, rising to the position of staff director for the majority party. After 1994, when the Republicans took control of the House, he became minority staff director and has continued to serve as a committee staff member up to the time of this writing. Other science fellows who stayed on as professional staff members include Benjamin S. Cooper, Michael L. Telson, and Leonard Weiss. For a complete list of the congressional fellows who have participated in the AAAS program—together with information on each individual's educational background, fellowship dates and assignment, sponsoring professional society, and current occupation—see DSPP (2000). Not all con-

gressional staffers specializing in science and technology policy launched their careers through AAAS's fellowship program, of course. Others who have been influential as congressional staff in science and technology policy include Radford Byerly, David Clement, David Goldston, John Holmfeld, Harold Hanson, Edith Hollerman, James Paul, Robert Roach, Beth Robinson, William Smith, William A. (Skip) Stiles Jr., Jim Turner, Harlan Watson, Tom Weimer, James Wilson, and Phillip Yeager.

[15]Congress has been wary of creating joint committees (apart from the analytic Joint Economic Committee and the Joint Committee on Printing, the Joint Committee on the Library, and the Joint Committee on Taxation) since the experience with the JCAE and has been alert to any undue concentration of technical expertise in a single body or committee. For example, when the Select Committee on Government Research—which had been created within the House of Representatives in 1963 and chaired by Carl Y. Elliott (Democrat, Alabama)—recommended that Congress create a Joint Committee on Research Policy "as a counterpart to Office of Science and Technology, the Federal Council on Science and Technology, and the President's Science Advisory Committee complex," the proposal went nowhere. See Select Committee on Government Research (1964, 55).

[16]For succinct discussions of various S&T issues addressed by the Senate up to this time, see *A Brief History of the Senate Committee on Commerce, Science, and Transportation* (1978).

[17]Among these first members were Lee A. DuBridge, W. Albert Noyes Jr., Roger Revelle, H. Guyford Stever, James A. Van Allen, and Fred L. Whipple.

[18]In 1963, the Science, Research and Development Subcommittee created its own Research Management Advisory Panel, which was meant to provide subcommittee members with informal advice on problems within applied science research programs that merited congressional intervention. This experimental advisory apparatus was eventually eclipsed by the subcommittee's more formal hearings and by information provided by the Legislative Reference Service (later the Congressional Research Service). See Hechler (1980, 135–136); and Stine (1986, 50–52).

[19]For an overview of the scientific and technical advice provided by these four agencies, see Committee on Science, Technology, and Congress (1991a). The individual agencies are analyzed in detail in Bimber (1990); Carroll (1990); Ellwood (1990); Eschwege (1990); Nichols (1990); and Marzotto (1991). For a summary of how the support agencies have been used by congressional members and their key staff, as well as how their roles could be improved, see Thurber (1981).

[20]For subsequent developments in information technologies and needs within the legislative branch, see Frantzich (1982); Chartrand (1986); Griffith (1991); and Chartrand and Ketcham (1993). Since the 104th Congress, CRS has been making its products and services available exclusively to members of Congress and their staff electronically via the congressional intranet at www.loc.gov/crs. See CRS (2001, 34).

[21]Representative Emilio Q. Daddario (Democrat, Connecticut), who chaired the Subcommittee on Science, Research, and Development in 1966, claimed that the Science Policy Research Division gave "Congress an ability which it didn't have prior to this time and has helped to increase the confidence of the Congress in its ability to handle matters of highly complicated scientific and technical involvement" (Daddario 1967, 116). See also Subcommittee on Science, Research, and Development (1964).

[22]See also Anderson (1964); Lowe (1965); and Frye (1966).

[23]CRS staff dropped to 838 in 1992 and gradually fell to 691 in 2001 (information provided by CRS staff in June 2001). See also Mosher (1979, 271).

[24]Edward Wenk Jr. directed the Science Policy Research Division from 1964 to 1966. (He had served as the Legislative Reference Service's first senior specialist in science and technology from 1959 to 1961.) The division's other professional staff gradually built up to include such analysts as Charles Sheldon, Robert Chartrand, Genevieve Knezo, Walter Hahn, and Richard Rowberg. Huddle operated semiautonomously, as was the privilege of senior specialists throughout the LRS (and later the CRS). When it reorganized itself in the late 1990s, CRS assigned all of its senior specialists to specific divisions, rather than allowing them to exist in an independent organizational unit of their own.

[25]For an account of these developments and the effect on the State Department, see Smith (1992, 137–154).

[26]For a thorough description of CRS's organization, staffing, and current activities, see CRS (2001).

[27]GAO was also charged with assisting individual members of Congress, as well as committees. CBO and OTA, on the other hand, were restricted to supporting only committees.

[28]For the history of the General Accounting Office, see Mansfield (1939); Pois (1979); and Mosher (1979, 1984).

[29]The GAO staff went from some 15,000 to 5,000 within several years after the war and remained at roughly this size until the early 1990s.

[30]According to the Committee on Science, Technology, and Congress (1991a, 32), GAO management had downgraded S&T policy to an "area of research" because it had "required less than 2.5 staff-years of effort" annually. The continuing minor usage of S&T staff resources eventually led GAO management to drop the category entirely as one of the agency's specialties.

[31]Although the CBO tends not to provide scientific and technical advice per se, Congress has occasionally called upon it to provide economic analyses of current or proposed science and technology policies. See, for example, CBO (1988a, 1988b, 1988c, 1990, 1991).

[32]For a discussion and suggestion of proposals of such steps that do not involve creating new organizations or formal arrangements, see Stine and LaFollette (1990).

[33]The NAS is a private, nonprofit organization chartered by Congress on March 3, 1863, to provide, on request, scientific advice to any department of government with science-related questions. The investigations, examinations, experiments, and reports produced by the NAS for executive branch agencies were paid for via congressional appropriations. In 1916, as war raged in Europe, the NAS established the National Research Council, which enabled it to expand its governmental consultative role by enlisting the voluntary services of the nation's most prominent scientists and engineers. The NAS expanded its complex through the creation of the National Academy of Engineering in 1964 and the Institute of Medicine in 1970. See Cochrane (1978) and Dupree (1979).

[34]One significant study, *Basic Research and National Goals*, which emerged from this arrangement with the House subcommittee, attracted a wide audience. The volume was a collection of essays dealing with the establishment of priorities in the public funding of scientific research. These priorities were extensively debated in the scientific community and among policy analysts. The controversy helped trigger a debate on "criteria for scientific choice," which eventually included a significant literature in the science affairs field (NAS 1965). For a review of the NAS report and related literature, see Smith (1966). Chairman Emilio Daddario later claimed that his subcommittee's 1964

contract with NAS proved to be "the first formal relationship that the National Academy had had with the Congress" (Daddario 1967, 115).

[35]For a list of congressionally mandated studies, see the NAS's website at nationalacademies.org (accessed May 8, 2003). For earlier NAS studies, see Committee on Science and Technology (1986).

[36]The Federal Advisory Committee Act Amendments of 1997, Public Law 105–153, December 17, 1997, created a new Section 15 of the FACA applying to the NAS and to the National Academy of Public Administration, in which these two organizations, as not-for-profit, nongovernmental entities, were exempted from certain FACA requirements, and in which Congress removed the legal ambiguities arising from *Animal Legal Defense Fund* v. *Shalala*, 63 F. 3d 383 (D.C. Cir. 1995); *Animal Legal Defense Fund* v. *Shalala*, 104 F. 3d 424 (D.C. Cir. 1997), certiorari denied by the U.S. Supreme Court for this and a companion case, *National Resources Defense Council* v. *Pena*, 118 S. Ct. 364 (1997); and *National Resources Defense Council* v. *Pena*, 147 F. 3d 1012 (D.C. Cir. 1998). For an earlier critical assessment of NAS, see Boffey (1975).

[37]For perceptive treatments of OTA, see Skolnikoff (1976); Committee on Science and Technology (1978); Gibbons and Gwin (1988); Carson (1992); Kunkle (1995); and Bimber (1996). For the background of the technology assessment idea, see Pursell (1974).

[38]For an autobiographical account of his OTA directorship, see the relevant sections in Peterson (1999).

[39]For a detailed examination of OTA studies and operations, see Guston (2001). A full list of OTA studies may be found at www.wws.princeton.edu/~ota/ (accessed May 8, 2003).

[40]For the political controversy surrounding this issue, see Graham (1992).

[41]This paradox, which was made evident through OTA's work on industrial policy, was likewise seen in the reactions to the agency's negative assessments of the Reagan administration's proposals for antiballistic missile defense systems. See, for example, OTA (1985a, 1985b, 1988). The perception of OTA's partisan identification was most recently discussed in "Time for a Bipartisan OTA" (2001) and Greenberg (2001, 288–290).

[42]President Clinton's appointment of Jack Gibbons as head of the White House Office of Science and Technology Policy and as Assistant to the President for Science and Technology confirmed the suspicions of some OTA critics, who had long believed that the agency's early political biases had never fully disappeared. Whereas Gibbons had proven himself a competent scientist and administrator, his leap straight from OTA to the Office of Science and Technology Policy was seen—rightly or wrongly—as confirmation that he was, at heart, a partisan Democrat. See Gibbons (1997) for material on his science advising to both Congress and the White House.

[43]The few published outcries that did appear were exemplified by Morgan (1995a, 1995b).

[44]For a fuller discussion of the OTA concept, see Epstein and Carter (2001). For a discussion of alternative measures to a formal OTA, including potential efforts to strengthen other congressional support agencies, see Hill (2001).

[45]The Ehlers Report (Committee on Science 1998) called on the scientific community to encourage its members to active roles in advocacy and public debate of issues. For an earlier proposal that scientists form voluntary, confidential, informal advisory panels that are locally based to serve each interested member of Congress, see Katz (1988).

[46]The Brookings Institution in 1981–1983 ran a science fellows program under a $1 million grant from the Alfred P. Sloan Foundation, which brought scientists to its staff

for a one-year period to "retool" and to pursue policy analysis projects of their own choosing. The program lapsed when the funding did. Significant titles that came from this project include Lave and Omenn (1981); Morgan (1984); Panem (1984); and Panem (1988). See also Task Force on Nongovernmental Organizations in Science and Technology (1993).

[47]Through funding from the Carnegie Commission on Science, Technology, and Government, the AAAS's Committee on Science, Engineering, and Public Policy produced a working manual aimed at improving communication between scientists and the legislative branch on policy matters relating to science and technology. See Wells (1992).

References

Abelson, Philip H. 1988. Scientific Advice to the Congress. In *Science and Technology Advice to the President, Congress, and Judiciary,* edited by William T. Golden. New York: Pergamon Press, 395–399.

Aberbach, Joel D. 1990. *Keeping a Watchful Eye: The Politics of Congressional Oversight*. Washington, DC: Brookings Institution.

Anderson, Clinton P. 1964. Scientific Advice for Congress. *Science* 144(April 3): 29–32.

Bimber, Bruce. 1990. Congressional Support Agency Products and Services for Science and Technology Issues: A Survey of Congressional Staff Attitudes about the Work of CBO, CRS, GAO, and OTA. Paper prepared for the Carnegie Commission on Science, Technology, and Government. New York.

———. 1996. *The Politics of Expertise in Congress: The Rise and Fall of the Office of Technology Assessment*. Albany: State University of New York Press.

Boffey, Philip M. 1975. *The Brain Bank of America: An Inquiry into the Politics of Science*. New York: McGraw-Hill.

A Brief History of the Senate Committee on Commerce, Science, and Transportation. 1978. Senate Doc. No. 95–93. 95th Cong., 2d sess. Washington, DC: U.S. Government Printing Office.

Carroll, James D. 1990. New Directions for the Congressional Research Service on Science and Technology Issues. Background report prepared for the Committee on Science, Technology, and Congress, Carnegie Commission on Science, Technology, and Government, New York.

Carson, Nancy. 1992. Process, Prescience, and Pragmatism: The Office of Technology Assessment. In *Organizations for Policy Analysis: Helping Government Think*, edited by Carol H. Weiss. Newbury Park, CA: Sage Publications, 236–251.

Chartrand, Robert Lee. 1986. Information Technology in the Legislative Process, 1976–1985. *Annual Review of Information Science and Technology* 21: 203–239.

Chartrand, Robert Lee, and Robert C. Ketcham. 1993. *Opportunities for the Use of Information Resources and Advanced Technologies in Congress: A Study*. New York: Carnegie Commission on Science, Technology, and Government.

Cochrane, Rexmond Canning. 1978. *The National Academy of Sciences: The First Hundred Years, 1863–1963*. Washington, DC: National Academy of Sciences.

Committee on Science. 1998. *Unlocking Our Future: Toward a New National Science Policy*. Report to the U.S. House of Representatives, 105th Cong., 2d sess. Washington, DC: U.S. Government Printing Office. (This report to Congress is informally known as the "Ehlers Report.")

Committee on Science and Technology. 1978. *Review of the Office of Technology Assessment and Its Organic Act*. Subcommittee on Science, Research and Technology, U.S. House of Representatives, 95th Cong., 2d sess. Washington, DC: U.S. Government Printing Office.

———. 1981. *Survey of Science and Technology Issues, Present and Future*. Staff report to the U.S. House of Representatives, 97th Cong., 1st sess. Washington, DC: U.S. Government Printing Office.

———. 1986. *Bibliography of Reports by the National Academy of Sciences, 1945–1985*. Science Policy Study Background Report No. 2, Part B, Task Force on Science Policy, U.S. House of Representatives, 99th Cong., 2d sess. Washington, DC: U.S. Government Printing Office.

Committee on Science, Technology, and Congress. 1991a. *Science, Technology, and Congress: Analysis and Advice from the Congressional Support Agencies*. New York: Carnegie Commission on Science, Technology, and Government.

———. 1991b. *Science, Technology, and Congress: Expert Advice and the Decision-Making Process*. New York: Carnegie Commission on Science, Technology, and Government.

———. 1994. *Science, Technology, and Congress: Organizational and Procedural Reforms*. New York: Carnegie Commission on Science, Technology, and Government.

Congressional Budget Office (CBO). 1988a. *Using Federal R&D To Promote Commercial Innovation*. Washington, DC: U.S. Government Printing Office.

———. 1988b. *Risks and Benefits of Building the Superconducting Super Collider*. Washington, DC: U.S. Government Printing Office.

———. 1988c. *The NASA Program in the 1990s and Beyond*. Washington, DC: U.S. Government Printing Office.

———. 1990. *Using R&D Consortia for Commercial Innovation: SEMATECH, X-ray Lithography, and High-Resolution Systems*. Washington, DC: U.S. Government Printing Office.

———. 1991. *Encouraging Private Investment in Space Activities*. Washington, DC: U.S. Government Printing Office.

Congressional Research Service (CRS). 1971. *Technical Information for Congress*. Science Policy Research Division, Library of Congress. Report to the Subcommittee on Science, Research and Development, Committee on Science and Astronautics, U.S. House of Representatives, 92nd Cong., 1st sess. Washington, DC: U.S. Government Printing Office.

———. 2001. *Annual Report of the Congressional Research Service of the Library of Congress for Fiscal Year 2000*. Submitted to the Joint Committee on the Library, U.S. Congress. Washington, DC: Congressional Research Service.

Daddario, Emilio. 1967. Science and the Congress. In *Washington Colloquium on Science and Society*, edited by Morton Leeds. Second series. Baltimore: Mono Book Corp.

Deering, Christopher J., and Steven S. Smith. 1985. Subcommittees in Congress. In *Congress Reconsidered*, 3rd ed., edited by Lawrence C. Dodd and Bruce I. Oppenheimer. Washington, DC: CQ Press, 189–210.

Directorate for Science and Policy Programs (DSPP). 2000. *Directory of AAAS Science and Engineering Fellows, 1973–2000*. Washington, DC: American Association for the Advancement of Science.

Dupree, A. Hunter. 1979. The National Academy of Sciences and the American Definition of Science. In *The Organization of Science in Modern America, 1860–1920*, edited by Alexandra Oleson and John Voss. Baltimore: Johns Hopkins University Press, 342–363.

Dyke, Richard Wayne. 1989. *Mr. Atomic Energy: Congressman Chet Holifield and Atomic Energy Affairs, 1945–1974*. New York: Greenwood.

Ellwood, John W. 1990. The Congressional Budget Office and the Improvement of Congress's Ability To Handle Expert and Science Information. Background report prepared for the Committee on Science, Technology, and Congress, Carnegie Commission on Science, Technology, and Government, New York.

Epstein, Gerald L., and Ashton B. Carter. 2001. Model 3: Establishing a Dedicated Support Agency To Provide Congress with Scientific and Technological Advice. Paper presented at the Workshop on Science and Technology Advice to Congress. June 14, 2001, Washington, DC.

Eschwege, Henry. 1990. Analysis of Science and Technology Issues for the Congress: Future Directions for the General Accounting Office. Background report prepared for the Committee on Science, Technology, and Congress, Carnegie Commission on Science, Technology, and Government, New York.

Foreman, Christopher H., Jr. 1988. *Signals from the Hill: Congressional Oversight and the Challenge of Social Regulation*. New Haven: Yale University Press.

Fox, Harrison W., Jr., and Susan W. Hammond. 1977. *Congressional Staffs: The Invisible Force in American Lawmaking*. New York: Free Press.

Frantzich, Stephen E. 1982. *Computers in Congress: The Politics of Information*. Beverly Hills, CA: Sage Publications.

Frye, Alton. 1966. *The Legislative Role in Science Policy: Congressional Perspectives and Mechanisms*. Los Angeles: Institute of Government and Public Affairs, University of California.

Gibbons, John H. 1997. *This Gifted Age: Science and Technology at the Millennium*. New York: Springer-Verlag Press.

Gibbons, John H., and Holly L. Gwin. 1988. Technology and Governance: The Development of the Office of Technology Assessment. In *Technology and Politics*, edited by Michael E. Kraft and Norman J. Vig. Durham, NC: Duke University Press, 98–122.

Gillespie, Ed, and Bob Schellhas (eds.). 1994. *Contract with America: The Bold Plan by Rep. Newt Gingrich, Rep. Dick Armey and the House Republicans To Change the Nation*. New York: Times Books.

Graham, Otis L., Jr. 1992. *Losing Time: The Industrial Policy Debate*. Cambridge, MA: Harvard University Press.

Green, Harold P., and Alan Rosenthal. 1961. *The Joint Committee on Atomic Energy: A Study in Fusion of Government Power*. Washington, DC: George Washington University.

Greenberg, Daniel S. 2001. *Science, Money, and Politics: Political Triumph and Ethical Erosion*. Chicago: University of Chicago Press.

Griffith, Jeffrey C. 1991. The Development of Information Technology in the Congressional Research Service of the Library of Congress. *Government Information Quarterly* 8(3): 293–307.

Guston, David H. 2001. Science and Technology Advice for the Congress: Insights from the OTA Experience. Paper presented at the Workshop on Science and Technology Advice to Congress. June 14, 2001, Washington, DC.

Hammond, Susan Webb. 1994. Congressional Staffs. In *Encyclopedia of the American Legislative System*, edited by Joel H. Silbey. New York: Charles Scribner's Sons, vol. II, 785–800.

Hechler, Ken. 1980. *Toward the Endless Frontier: History of the Committee on Science and Technology, 1959–79*. Washington, DC: U.S. Government Printing Office.

Heller, Edward. 1991. Interview by Jeffrey K. Stine. November 13 and December 4, 1991. Engineering Collections, National Museum of American History, Smithsonian Institution, Washington, DC.

Henschen, Beth M., and Edward I. Sidlow. 1986. The Recruitment and Career Patterns of Congressional Committee Staffs: An Exploration. *Western Political Quarterly* 39 (December): 701–708.

Hill, Christopher T. 2001. Model 2: Create a New Science and Technology Policy Advisory Unit To Be Located within an Existing Legislative Branch Agency. Paper presented at the Workshop on Science and Technology Advice to Congress. June 14, 2001, Washington, DC.

Katz, James Everett. 1988. Congress Needs Informal Science Advisors: A Proposal for a New Advisory Mechanism. In *Science and Technology Advice to the President, Congress, and Judiciary*, edited by William T. Golden. New York: Pergamon Press, 425–430.

Kofmehl, Kenneth. 1977. *Professional Staffs of Congress*, 3rd ed. West Lafayette, IN: Purdue University Press.

Kunkle, Gregory C. 1995. New Challenge or the Past Revisited? The Office of Technology Assessment in Historical Context. *Technology in Society* 17(2): 175–196.

LaFollette, Marcel C. 1990. Congressional Oversight of Science and Technology Programs. Report prepared for the Committee on Science, Technology, and Congress, Carnegie Commission on Science, Technology, and Government, New York.

Lakoff, Sanford A. 1974. Congress and National Science Policy. *Political Science Quarterly* 89(Fall): 589–611.

Lave, Lester B., and Gilbert S. Omenn. 1981. *Cleaning the Air: Reforming the Clean Air Act*. Washington, DC: Brookings Institution.

Lowe, George E. 1965. Congress and Science Advice. *Bulletin of the Atomic Scientists* 21(December): 39–42.

Maddox, Robert F. 1979. The Politics of World War II Science: Senator Harley M. Kilgore and the Legislative Origins of the National Science Foundation. *West Virginia History* 41(Fall): 20–39.

Malbin, Michael J. 1980. *Unelected Representatives: Congressional Staff and the Future of Representative Government*. New York: Basic Books.

Mansfield, Harvey C. 1939. *The Comptroller General: A Study in the Law and Practice of Financial Administration*. New Haven: Yale University Press.

Marzotto, Antonette. 1991. Recruitment and Retention of Senior Science and Technology Personnel in the Congressional Research Service and Office of Technology Assessment. Background report prepared for the Committee on Science, Technology, and Congress, Carnegie Commission on Science, Technology, and Government, New York.

Meltsner, Arnold J. 1990. *Rules for Rulers: The Politics of Advice*. Philadelphia: Temple University Press.

Morgan, M. Granger. 1995a. The Office of Technology Assessment: An Endangered Species Worth Saving. *IEEE Spectrum* 35(February): 13–14.

———. 1995b. Death by Congressional Ignorance. *Pittsburgh Post–Gazette*, August 2.

Morgan, Robert P. 1984. *Science and Technology for International Development: An Assessment of U.S. Policies and Programs*. Boulder, CO: Westview Press.

Mosher, Frederick C. 1979. *The GAO: The Quest for Accountability in American Government*. Boulder, CO: Westview Press.

———. 1984. *A Tale of Two Agencies: A Comparative Analysis of the General Accounting Office and the Office of Management and Budget*. Baton Rouge: Louisiana State University Press.

Mumford, L. Quincy. 1958. Statement. In Science and Technology Act of 1958, hearings before the Subcommittee on Reorganization, Committee on Government Operations, U.S. Senate, 85th Cong., 2d sess. Washington, DC: U.S. Government Printing Office, part 2, 367.

National Academy of Sciences (NAS). 1965. *Basic Research and National Goals*. Report to the Committee on Science and Astronautics, U.S. House of Representatives. Washington, DC: U.S. Government Printing Office.

Nichols, Rodney W. 1990. Vital Signs OK: On the Future Directions of the Office of Technology Assessment. Background report prepared for the Committee on Science, Technology, and Congress, Carnegie Commission on Science, Technology, and Government, New York.

Office of Technology Assessment (OTA). 1985a. *Ballistic Missile Defense Technologies*. Washington, DC: U.S. Government Printing Office.

———. 1985b. *Anti-Satellite Weapons, Countermeasures, and Arms Control*. Washington, DC: U.S. Government Printing Office.

———. 1986. *The Regulatory Environment for Science*. Science Policy Study Background Report No. 10. Task Force on Science Policy, Committee on Science and Technology, U.S. House of Representatives, 99th Cong., 2d sess. Washington, DC: U.S. Government Printing Office.

———. 1988. *SDI: Technology, Survivability, and Software*. Washington, DC: U.S. Government Printing Office.

Ogul, Morris S. 1976. *Congress Oversees the Bureaucracy: Studies in Legislative Supervision*. Pittsburgh: University of Pittsburgh Press.

Orlans, Harold (ed.). 1968. *Science Policy and the University*. Washington, DC: Brookings Institution.

Ornstein, Norman J., Thomas E. Mann, and Michael J. Malbin. 2000. *Vital Statistics on Congress, 1999–2000*. Washington, DC: AEI Press.

Panem, Sandra. 1984. *The Interferon Crusade*. Washington, DC: Brookings Institution.

———. 1988. *The AIDS Bureaucracy*. Cambridge, MA: Harvard University Press.

Peterson, Russell W. 1999. *Rebel with a Conscience*. Newark: University of Delaware Press.

Pois, Joseph. 1979. *Watchdog on the Potomac: A Study of the Comptroller General of the United States*. Washington, DC: University Press of America.

Pursell, Carroll. 1974. Belling the Cat: A Critique of Technology Assessment. *Lex et Scientia* 10(October–December): 130–145.

Rundquist, Paul S., Judy Schneider, and Frederick H. Pauls. 1992. *Congressional Staff: An Analysis of Their Roles, Functions, and Impacts*. Washington, DC: Congressional Research Service, Library of Congress.

Salisbury, Robert H., and Kenneth A. Shepsle. 1981. Congressional Staff Turnover and the Ties-That-Bind. *American Political Science Review* 75(June): 381–396.

Science and the Congress: The Third Franklin Conference. 1978. Philadelphia: Franklin Institute Press.

Select Committee on Government Research. 1964. *National Goals and Policies*. Report to U.S. House of Representatives, 88th Cong., 2d sess. Washington, DC: U.S. Government Printing Office.

Skolnikoff, Eugene B. 1976. The Office of Technology Assessment. In *Congressional Support Agencies: A Compilation of Papers*. Prepared for the Commission on the

Operation of the Senate, 94th Cong., 2d sess. Washington, DC: U.S. Government Printing Office, 55–74.

Smith, Bruce L.R. 1966. The Concept of Scientific Choice: A Brief Review of the Literature. *American Behavioral Scientist* 9(May): 27–36.

———. 1992. *The Advisers: Scientists in the Policy Process*. Washington, DC: Brookings Institution.

Smith, Marcia S., John D. Moteff, Lennard G. Kruger, Glenn J. McLoughlin, and Jeffrey W. Seifert. 2001. *Internet: An Overview of Key Technology Policy Issues Affecting Its Use and Growth*. CRS Report 98–67 STM, January 31. Washington, DC: Congressional Research Service.

Stine, Jeffrey K. 1986. *A History of Science Policy in the United States, 1940–1985*. Science Policy Study Background Report No. 1. Task Force on Science Policy, Committee on Science and Technology, U.S. House of Representatives, 99th Cong., 2d sess. Washington, DC: U.S. Government Printing Office.

———. 1988. Fulfilling the Science and Technology Advisory Needs of Congress. In *Science and Technology Advice to the President, Congress, and Judiciary,* edited by William T. Golden. New York: Pergamon Press, 444–445.

———. 1994. *Twenty Years of Science in the Public Interest: A History of the Congressional Science and Engineering Fellowship Program*. Washington, DC: American Association for the Advancement of Science.

Stine, Jeffrey K., and Marcel C. LaFollette. 1990. *Congressional Hearings on Science and Technology Issues: Strengths, Weaknesses, and Suggested Improvements*. Background report prepared for the Committee on Science, Technology, and Congress, Carnegie Commission on Science, Technology, and Government, New York.

Subcommittee on Science, Research, and Development. 1964. *Scientific and Technical Advice for the Congress: Needs and Sources*. Report to the U.S. House of Representatives, 88th Cong., 2d sess. Washington, DC: U.S. Government Printing Office.

Subcommittee on War Mobilization. 1945. *The Government's Wartime Research and Development, 1940–1944*. Committee on Military Affairs, U.S. Senate, 79th Cong., 1st sess. Washington, DC: U.S. Government Printing Office.

Task Force on Nongovernmental Organizations in Science and Technology. 1993. *Facing toward Governments: Nongovernmental Organizations and Scientific and Technical Advice*. New York: Carnegie Commission on Science, Technology, and Government.

Thurber, James A. 1981. The Evolving Role and Effectiveness of the Congressional Research Agencies. In *The House at Work,* edited by Joseph Cooper and G. Calvin Mackenzie. Austin: University of Texas Press, 292–315.

Time for a Bipartisan OTA. 2001. *Nature* 411(May 10): 117.

Vig, Norman J. 2003. The European Experience. In *Science and Technology Advice for Congress,* edited by M. Granger Morgan and Jon M. Peha. Washington, DC: Resources for the Future, Chapter 5.

Vig, Norman J., and Herbert Paschen (eds.). 2000. *Parliaments and Technology: The Development of Technology Assessment in Europe*. Albany: State University of New York Press.

Wells, William G., Jr. 1992. *Working with Congress: A Practical Guide for Scientists and Engineers*. Washington, DC: American Association for the Advancement of Science.

3

The Origins, Accomplishments, and Demise of the Office of Technology Assessment

Robert M. Margolis and David H. Guston

During the summer of 1995, after multiple rounds of debates, the U.S. Congress abolished one of its four support agencies: the Office of Technology Assessment (OTA). This unprecedented event was important, not because of the size of OTA's budget, but because of OTA's stature as a source of technically substantive, balanced analysis for Congress. In the ensuing years analysts and politicians alike have continued to debate how to best provide scientific and technical advice for use in the congressional decisionmaking process. As the remainder of this book will help to clarify, there are a number of institutional models that may be able to fill the void left by the closure of the OTA.

To better understand how these competing models may be able to serve the unique needs of Congress with respect to science and technology issues, it is useful to place OTA in historical context. OTA's origins were rooted in the turbulence of the 1960s and early 1970s. During this time, people began to question seriously the role of science and technology in society, government spending on research and development (R&D) was growing, legislative activities related to science and technology were increasing, and the executive branch was expanding and commanded large analytical resources not available to Congress. Largely in response to these issues, Congress passed legislation forming OTA in 1972.

During its early years, OTA encountered a number of obstacles. In particular, balancing its mandated role of serving Congress with the ideal of conducting independent technology assessments proved to be a formidable task. However, by the early 1980s OTA had developed an approach to technology assessment that met the needs of Congress, included a wide range of policy actors, and produced useful results. OTA appeared to have earned a permanent place in the congressional process. Yet in 1994 a substantial shift of power occurred in Congress. This shift of power led directly to the demise of OTA.

In the body of this chapter, we describe OTA's origins and evolution over time. We also discuss its accomplishments, in particular, how it raised the level of debate in Congress and influenced the legislative process, how it saved U.S. taxpayers money, and how it served democracy by bringing a wide range of views into many debates on science and technology issues. In addition, we examine how the shift of power in Congress in 1994 led to the demise of OTA.

The Origins of OTA

The seeds for OTA grew out of the turbulence of the 1960s and early 1970s. In particular, four key issues led to the creation of OTA in 1972: (1) dramatic growth in the U.S. government's science and technology budget, (2) increased questioning of the social and political benefits of science and technology in society, (3) an increasing share of legislative activities concerned with science and technology issues, and (4) rising concern about the balance of analytical resources and power between the executive and legislative branches of government.[1]

Increasing Spending on R&D

The U.S. government's spending on scientific research and development (R&D) began to rise during the early 1950s and accelerated rapidly in response to the launching of *Sputnik* in 1957. The U.S. government's R&D budget increased from $3 billion in the mid-1950s to $17 billion by 1972 (NSF 1993). Even in constant dollars, this represented a significant increase, as shown in Figure 3-1. In the year that OTA was finally created, 1972, President Nixon requested an increase of more than $1 billion in research spending for fiscal year 1973 (Leg. Hist. P.L. 92–484, 3578). Thus, over time Congress was being asked to allocate more and more money for science and technology, yet it had limited ability to internally evaluate these investments.

Questioning the Role of Science and Technology

Whereas science and technology have long been viewed as having both positive and negative effects on society, in the 1960s questions about the role of science and technology in society became widespread. In particular, a number of controversies surrounding the environment, nuclear weapons, the supersonic transport (SST), the use of defoliants in the Vietnam War, and other contentious issues led to increasing concerns about the role of science and technology in society. As von Hippel and Primack (1991) point out, during this period critical voices within the scientific community and the broader public, such as Rachel Carson, Ralph Nader, and Barry Commoner, helped to raise

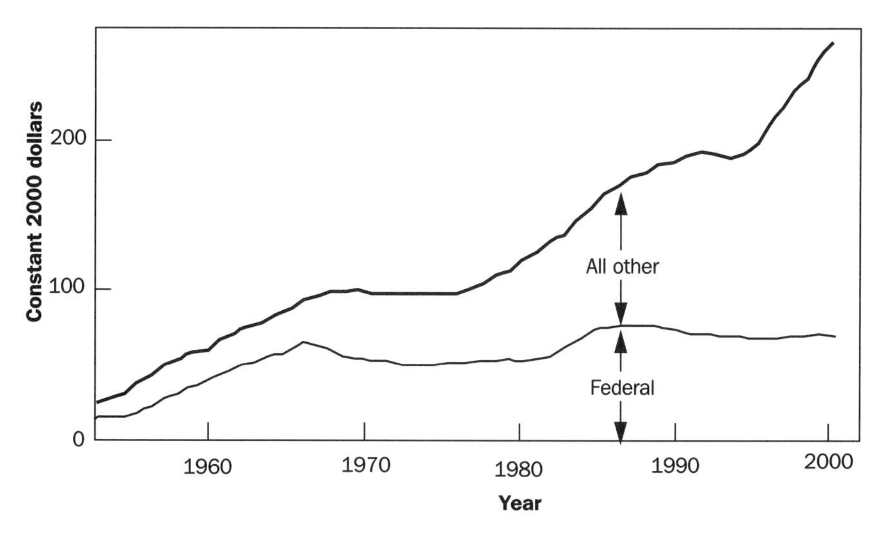

Figure 3-1. R&D Expenditures in the United States from 1953 through 2000 in 2000 Constant Dollars

Source: Adapted from NSF 2000.

public awareness and bring these types of issues into the policy arena. As Congress debated complex technological issues, some of its members recognized the need to institutionalize better means of addressing them.

As part of this process, in 1963, Congress established the Subcommittee on Science, Research, and Development under the chairmanship of Representative Emilio Daddario (Democrat, Connecticut). This event was important in leading up to the creation of OTA.[2] In many ways, Daddario captured the broad spirit of questioning the role of science and technology during the 1960s. Far from being a Luddite, Daddario viewed science and technology as essential for America's future. However, he also acknowledged that science and technology can have a darker side. As he observed in 1963, "[Congress] finds itself squarely faced with the many social, political, and economic side effects created by the current technological revolution" (quoted in Kunkle 1995, 179). By the mid-1960s, Daddario's subcommittee began to explore the possibility of establishing a technology assessment mechanism within Congress. This mechanism would be designed to look at both the desirable and undesirable consequences of science and technology. Eventually, in 1972 Daddario became the first director of OTA.

Increasing Legislative Activities Related to Science and Technology

Along with spending, oversight and legislative activities related to science and technology issues increased in Congress during the late 1960s. For example, between 1965 and 1972, 39 congressional committees or subcommittees and four joint committees or subcommittees engaged in important legislative

Table 3-1. Selected Legislative Activities Related to Technology Assessment (1965–1972)

Field	*Example issues*
Commerce	Supersonic transport (SST) Northeast corridor experiment Weather modification Desalination of seawater
Energy	Nuclear power development The Alaskan oil pipeline Hydroelectric power vs. ecology in the Northwest
Environmental	Trace metal poisons Food supplies Pesticides Antibiotic stock feed Development of electric automobile engines Strip-mining techniques
Miscellaneous	Space exploration Sea-bed mineral resources Antarctic investigation Global atmospheric research Many defense-related issues

Source: Leg. Hist. P.L. 92–484, 3574.

activities directly related to technology assessment (Leg. Hist. P.L. 92–484, 3574). As shown in Table 3-1, the legislative activities related to technology assessment dealt with in Congress during this period covered a wide range of fields and issues. Yet without an internal, unbiased source of technical information, Congress was struggling with the need to make increasingly complex decisions about science and technology issues.

In addition, during the early 1970s, increased activities within Congress related to public health and welfare helped to highlight further the need for independent congressional evaluation of science and technology issues. For example, in conducting an investigation into the effects of lead-based paint poisoning in 1972, Senator Edward Kennedy (Democrat, Massachusetts) pointed out that at the time about 400,000 children were suffering from lead poisoning and annually about 200 children were dying from the ingestion of 30-year-old lead-based paint. As Senator Kennedy observed,

> If Congress had had an Office of Technology Assessment 30 years ago, it is conceivable we could have anticipated this problem and enacted legislation which would have spared thousands of children from the grievous effects [of] this poison. (Leg. Hist. P.L. 92–484, 3579)

Advocates of establishing the OTA argued that it would be likely to help improve Congress's legislative and oversight functions with respect to a wide range of emerging technological issues, such as water management, long-distance pipelines, offshore airports, dissemination of medical care, ocean drilling, and the development of new sources of energy and materials (Leg. Hist. P.L. 92–484, 3587).

Balance of Power between the Executive and Legislative Branches

During the 1960s Congress became increasingly aware of a significant imbalance between the ability of the executive and legislative branches of government to analyze and evaluate technical issues. By the early 1970s, in particular under the Nixon administration, the executive branch had expanded considerably. This was especially true in terms of its ability to carry out assessments of science and technology issues. For example, by 1972 the Department of Agriculture, the Environmental Protection Agency, the Food and Drug Administration, the Defense Department, the Department of Transportation, and the Department of Health, Education, and Welfare had all sponsored or conducted at least partial technology assessments (Leg. Hist. P.L. 92–484, 3587).

Thus it is not surprising that by the early 1970s supporters of the idea of creating an independent technology assessment organization within Congress began to focus on the role that OTA would play in helping to restore balance between the executive and legislative branches of government. Looking again at Representative Daddario's role, we find that by 1970 he began to highlight this dimension of OTA. As he stated during hearings before the House Subcommittee on Science, Research and Development,

> We [Congress] have recognized the important need for developing independent means of obtaining necessary and relevant technical information for the Congress, without having to depend almost solely on the Executive Branch. In my view, it is only with this capability that Congress can assure its role as an equal branch in our Federal structure. (quoted in Kunkle 1995)

This power struggle between the executive and legislative branches of government went beyond science and technology issues. Thus Congress created OTA as part of a broader effort to strengthen its analytical support agencies during the early 1970s.[3] Creating OTA was a critical component of Congress's attempt to transform itself from being a passive recipient of external information on science and technology issues, particularly from the executive branch, into an active participant in determining how to frame and define science and technology issues.

Most important to recognize about OTA's founding for this chapter is that it was based on both an institutional (and not merely partisan) conflict of crisis proportions between Congress and the executive branch that implicated expertise and technology (as well as war powers and fiscal authority), and also by the intellectual and social movement of technology assessment that—

together with consumerism and environmentalism—upped the ante for the foresight involved in making public decisions.

The Difficult Early Years

After extensive hearings and studies, Congress passed legislation in 1972 to set up the OTA (Appendix 1). Section 2 of the Technology Assessment Act of 1972 clearly sets forth the assumptions and expectations of the new agency:

> (a) As technology continues to change and expand rapidly, its applications are—(1) large and growing in scale; and (2) increasingly extensive, pervasive, and critical in their impact, beneficial and adverse, on the natural and social environment.
>
> (b) Therefore, it is essential that, to the fullest extent possible, the consequences of technological applications be anticipated, understood, and considered in determination of public policy on existing and emerging national problems.
>
> (c) The Congress further finds that: (1) the Federal agencies presently responsible directly to the Congress are not designed to provide the Legislative Branch with adequate and timely information, independently developed, relating to the potential impact of technological applications, and (2) the present mechanisms of the Congress do not and are not designed to provide the Legislative Branch with such information.
>
> (d) Accordingly, it is necessary for the Congress to—(1) equip itself with new and effective means for securing competent, unbiased information concerning the physical, biological, economic, social, and political effects of such applications; and (2) utilize this information, whenever appropriate, as one factor in the legislative assessment of matters pending before the Congress, particularly in those instances where the Federal Government may be called upon to consider support for, or management or regulation of, technological applications. (P.L. 92–484, 928)

As we will see in a later section, transforming OTA from a concept into a functioning agency proved to be very challenging.

To govern OTA, the legislation established a Technology Assessment Board (TAB) consisting of six senators and six representatives, with equal numbers of Democrats and Republicans. The idea behind the TAB was to create a bipartisan, bicameral governing body for overseeing the office and its director—in essence it was designed to help OTA maintain political neutrality. Yet early on, under the direction of retired Representative Daddario, the office was highly politicized. As Weingarten points out, members of TAB appointed the staff and involved themselves closely in the assessments, essentially running the office like a congressional committee (Weingarten 1995).

The legislation also created an independent Technology Assessment Advisory Council (TAAC) to provide advice from experts outside of government.

TAAC was added during conference negotiations, largely in response to concerns from leaders in the science and technology community who were skeptical about the idea of technology assessment in general. TAAC consisted of 10 expert members of the public, appointed by the TAB, the comptroller general (who heads the General Accounting Office), and the director of the Congressional Research Service.

Whereas TAB had formal control over OTA's analytical agenda and remained engaged over OTA's history, TAAC had no formal operational authority and was, perhaps consequently, less active and engaged. Herdman and Jensen (1997) describe a change over time in the TAB from a kind of joint committee to a board of directors, and in TAAC from active managers to a visiting committee. From its inception, however, it was clear that TAAC would serve in a secondary role: it was not responsible for coming up with ideas for specific studies but rather was to serve as a sounding board for advice on topics coming out of congressional committees (Kunkle 1995). Thus from the beginning OTA focused primarily on serving congressional needs.

The needs of Congress as an institution, however, have often been in conflict with the desires of some (or many) of its individual members. As illustrated by the quote from OTA's enabling legislation, Congress initially wanted OTA to take a long-term view. In contrast, to maintain support from Congress, OTA had to balance these needs against the typical desires of individual members to focus on immediate issues and quick answers. Although OTA produced studies that were praised for their objectivity and balance, they were not nearly as far ranging as originally envisioned.

During its nascent years, OTA had access to the burgeoning literature on the philosophy and methods of technology assessment but paid it relatively little attention (Coates 1999), perhaps because the new office lacked a "critical mass of staff, resources, and experience to establish a consistent methodology—even at a general level" (Wood 1997). Toward the end of the term of Daddario, its first director (1973–1977), OTA made an initial effort to consolidate knowledge about methods of technology assessment in government and the private sector. The review, which culminated in a report based on hearings before the TAB (OTA 1977), concluded that technology assessment (1) was an increasingly useful tool for medium- and long-term management in the public and private sectors alike; (2) could provide early warning of unanticipated consequences as well as analysis of options and alternatives; and (3) should be "tailor-made to fit the resources, timing, and needs of the decision makers." The hearing highlighted research managers from industry who praised technology assessment as a managerial tool, in part perhaps to defuse some criticism from industry that greeted OTA's creation with fears of a regulatory body.[4]

After three and a half years as OTA's director, Daddario resigned in July 1977. At the time there was a flurry of controversy about OTA concerning the role of the TAAC and operational difficulties that OTA was experiencing (U.S. Congress 1978). In addition, there were accusations that OTA was being "taken over" by Democrats—in particular by Senator Edward Kennedy (Norman 1977). The agency's future was uncertain.

In January 1978, Russell Peterson, a Republican former governor of Delaware with strong credentials as an environmentalist, was appointed the second director of OTA. Peterson made a number of basic changes at OTA that helped clean up its internal affairs. For example, he won the right to have complete control over staff appointments. This helped return OTA to its non-partisan roots. However, he also tried to make OTA more independent from Congress. In a sense, Peterson wanted to move OTA toward fulfilling the early-warning and social impact assessment functions originally envisioned by Congress. As part of this effort, he initiated a priority-setting enterprise that solicited input from more than 5,000 members of the public (OTA 1979).

Wood (1997, 146) observes that this priority-setting process became politicized over concerns that OTA was "becoming too independent from congressional oversight and needs." Both Kunkle (1995) and Bimber (1996) concur, and the latter writes even more critically that "[t]he exercise was a classic policy analyst's attempt at determining national priorities through technical non-political means. It outraged many legislators who recognized it as a rejection of Congress's own agenda-setting processes." Thus under Peterson the relationship between OTA and Congress showed signs of shifting from the formal leadership of the TAB to the internal and informal sources of leadership within OTA (Coates 1982). Yet OTA's enabling legislation did not create OTA to be an independent think tank. Instead it created OTA to help Congress do its work related to science and technology issues.

Peterson left OTA after only a year and a half to become the president of the National Audubon Society, something that he had always wanted to do. Although there is no evidence that he was forced out, he did leave the young agency in an embattled state. The senior staff members who participated in Peterson's priority-setting process nevertheless devised criteria (which might be useful for posterity) for determining whether OTA might fruitfully conduct an assessment on any given topic (OTA 1979):

- Does the assessment involve the impact of technology?
- Is there congressional interest?
- Does the technology impact significantly on human needs and the quality of life?
- Would the assessment provide foresight?
- Can OTA do the assessment?

The Maturing of OTA

After Peterson's departure, John Gibbons, a former physicist at Oak Ridge National Laboratory and an energy and environmental policy specialist at the University of Tennessee, was appointed director of OTA. When he joined OTA in 1979, the agency's future was uncertain. One TAB member welcomed Gibbons by slapping him on the back, wishing him luck, and telling him that he was OTA's "last chance" (van Dam and Howard 1988). Yet Gibbons quickly turned the troubled agency around. He immediately moved away from Peterson's idea of making the agency more independent from Congress. Instead, he made it clear that under his direction OTA would be a nonpartisan servant of Congress. During his tenure at OTA, Gibbons—whose party affiliation was unknown when he was appointed—maintained a strict and distant neutrality with respect to partisan and jurisdictional maneuvering (Bimber 1996).

Gibbons held down the size of his staff, completed all ongoing studies, and initiated only tightly bound studies in response to congressional requests (Coates 1982). Thus, during his first year in office Gibbons instituted an "annual survey of committee needs" as he attempted to secure bipartisan support for studies that would be clearly tied to the legislative agenda of Congress.

The typical full assessment followed the 11 steps outlined in Box 3-1.[5] The "OTA process," as it was known generally, could not commence without a specific charge from Congress, but assessments were often stimulated by discussions among congressional and OTA staff and even informal solicitations from OTA. Committee chairs could request reports for themselves or on

Box 3-1
Stylized OTA Process

1. OTA staff have prerequest conversations with committee members and staff.
2. Committee(s) makes formal request of OTA for a study.
3. OTA submits project proposal to TAB.
4. TAB approves proposal.
5. OTA organizes staff and selects advisory panel.
6. OTA staff plan project and engage in data collection and analysis (including advisory panel meetings, workshops, contractor reports, briefings, surveys, site visits, etc.).
7. OTA staff draft final report (with revisions after both in-house review and external peer review).
8. OTA transmits draft to TAB for approval.
9. TAB approves and releases summary and full report (including embargoed press packet and press conference, electronic dissemination, and mailings to Congress, study participants, interested parties, and libraries).
10. OTA staff conducts policy outreach (including testimony at congressional hearings, briefings and informal talks with committee members and staff, interaction with staff of executive branch agencies, and addresses to various communities).
11. OTA staff pursue possible follow-on activities (such as the preparation of supporting documents, provision of more congressional testimony, and requesting new assessment activities).

Source: Derived from OTA Assessment Process (OTA 1995c).

behalf of the ranking minority member or a majority of the committee. The TAB and the OTA director, who was a nonvoting member of the TAB, could also request assessments. The TAB had to approve every proposal for an assessment before work began, helping to insulate the agenda from politicization by partisan interests or capture by individual committee agendas.

A staff of two to six analysts—including contractors—would then organize an advisory panel of (usually) nongovernmental experts and stakeholders to help scope, frame, and guide the assessment. Staff would pursue the assessment through a variety of methods, circulating preliminary drafts to the members of the advisory panel and, often, to additional outside readers. The final draft was subject to more formal internal and external review before being submitted to the director and the TAB for approval and release. Again, the bipartisan TAB as primary audience applied a strong discipline in the writing of reports. Congressional testimony and contact with administration officials, press, and stakeholder and public groups often followed the issuing of reports. In addition to publishing full reports, OTA also published smaller documents, briefed congressional staff, fielded inquiries, and provided testimony on an ad hoc basis.

In performing this work, OTA is variously described as having provided science advice (or scientific and technical advice) to Congress, as having conducted technology assessments (per its name), and as having performed policy analysis, particularly for issues with a high scientific or technological content. All three descriptions are accurate, perhaps to different degrees, and one could if so motivated sort OTA's written work into each of the categories. However, these three categories are themselves overly broad. Box 1-1 in Chapter 1 of this book presents a further classification of tasks in providing scientific and technical analysis. Although the table lists contemporary issues for each of the categories, OTA's own reports could easily be sorted into these same categories.

Under Gibbons, OTA continued its tradition of being reflexive about its work, initiating several large-scale internal studies and many more smaller discussions (Wood 1997). The first such study, the Task Force on TA [Technology Assessment] Methodology and Management, began shortly after Gibbons took office and was completed in 1980. This study's report crystallized consensus around the emerging OTA process, particularly the use of diverse methods—including advisory panels, workshops, and stakeholder participation—and the central role of staff. Earlier, OTA had made heavy use of contractors and, as is the case with reports from the National Academy of Sciences complex, relied on advisory panels for a great deal of the writing as well. Wood (1997), who chaired the task force, reports that it also demonstrated a consensus around the need for tighter management of OTA studies, including so-called "project review checkpoints" that would help ensure both timely completion and balanced, high-quality results, but that it did not achieve any consensus around "a deeper level of technology assessment methodology, nor on specific methods or techniques."

In September 1992, OTA began another self-study process to scrutinize and improve the work it conducted. This self-study marked a break from the past, identifying OTA's work as a specific form of policy analysis, although the printed report begged the question of what policy analysis is by defining it as the activity of policy analysts (OTA 1993). This operational sleight-of-hand, however, was not new for OTA, as technology assessment had often been defined not by a suite of techniques or intellectual perspectives but as whatever OTA happened to be doing. It was clear, though, that OTA's mission had metamorphosed from the early-warning aspect emphasized in its organic legislation to the provision of "thorough, objective information and analysis to help Members of Congress understand and plan for the short- and long-term consequences of the applications of technology, broadly defined" (OTA 1993).

The self-study identified two standard aspects of OTA's policy analysis: first, the description of the context of a policy problem and the presentation of the relevant issues or findings that might require congressional attention, and second, the discussion of potential solutions or options that Congress might choose to adopt. The appropriate balance of attention to context and options was not clear. OTA's particular brand of policy analysis was distinguished, of course, by "highlighting the relevant aspects of science and technology" and by its broad involvement of stakeholders in the process of analysis (OTA 1993).

From a process that included written evaluations of OTA reports from former congressional staff, telephone interviews with then-current congressional staff, and a workshop with 10 outside experts from different fields but familiar with OTA, the self-study identified three primary criteria of good policy analyses: objectivity, reader-friendliness, and timeliness. Congressional staff identified OTA's reputation for objectivity as one of its "chief assets" (OTA 1993). However, the study found that what staffers meant by objectivity seemed to vary from a lack of issue-related bias to evidence of using scientifically based literature and data. When presented with a selection of reports to evaluate for objectivity, the staffers found only minor departures from objectivity in a minority of the sample, but one report was severely criticized. The objections lodged against these reports centered on the apparent lack of empirical justification for some of the findings and the presentation of options that bordered too closely on, or lapsed into, recommendations. The staffers, however, offered no charges of partisanship.

Although objectivity may have been the primary desideratum for an OTA report, qualities not directly related to the analysis were also critical to OTA's congressional client. The self-study found that "reader-friendliness" and "timeliness" ranked with objectivity as the most important qualities in OTA reports (OTA 1993). The study found OTA's scores on these criteria a bit lower, as reports often lacked such reader-friendly production qualities as a useful executive summary (written with a structure parallel to the report) and a thorough index, and they often took two years or longer to produce (the higher the demands for objectivity and production quality, the longer production is likely to take). The importance of these criteria suggest that planners of a new institution should not overlook aspects of service to the congressional client that are ancillary to the intellectual performance of analysis itself but would require strict oversight to secure, particularly if a distributed system performed the analyses.

Overall, the self-study concluded that the quality of OTA's policy analysis was "often good—and frequently regarded as better than that of other policy organizations," but with "considerable variation in the quality and methods of policy analysis from report to report" (OTA 1993).

Most observers would agree that OTA experienced its "golden age" during Gibbons's 14-year tenure. A number of factors helped make OTA successful during this period: it strove to serve Congressional needs, it developed a process that incorporated a wide range of views into its analyses, and it provided a direct link to Congress for those outside the Beltway. During Gibbons's tenure, OTA produced a steady stream of high-quality, largely well-received reports that, as will be discussed in a later section, made constructive contributions to the legislative process.

In 1995, the year of its closure, OTA's budget represented about 1% of the legislative appropriations bill. The number of its staff hovered around 200, but because a significant number were contractors (not to mention fellows and detailees [employees of executive agencies, temporarily detailed to work for Congress]) who were with OTA only for a limited duration, the specific number of full-time OTA employees was often difficult to determine. Most employees were analysts with advanced degrees, working in a relatively flat organizational structure. OTA (1993) reported that 25% of the staff held a master's degree, 37% a Ph.D., and 10% a J.D. or M.D. Natural science and engineering accounted for 55% of the Ph.D.s and 42% of the master's degrees. The full-time employees contributed more general expertise, institutional memory, and specific knowledge of the congressional client; the contractors and others brought more specific expertise and links to external, ad hoc networks.

Despite its success, during the 1980s and early 1990s, problems continued to arise in a number of areas between OTA and Congress. Two areas in particular stand out: the timing of OTA reports and OTA's inability to make policy recommendations. As mentioned above, members of Congress are often focused on short-term issues and desire immediate answers to current questions, yet OTA was designed to be forward-looking and to take a long-term perspective on science and technology issues. OTA studies typically required a considerable amount of time to organize and conduct, on average about two years from conception to completion. This meant that some OTA reports, which from a political and technical perspective incorporated solid analysis, were completed after an issue had already faded from the political agenda.

In addition, the drafters of OTA's enabling legislation realized that to survive politically OTA would have to remain as nonpartisan as possible. Thus they specified that OTA should refrain from making policy recommendations; its role would be to clarify alternatives. This limited role for OTA was clearly stated in the legislative history of the OTA's enabling legislation, "The OTA would not be empowered to make recommendations for legislative action, or to render advice on courses of action, nor would it possess any kind of regulatory powers" (Leg. Hist. P.L. 92–484, 3584). Whereas this limitation helped OTA to gain the respect of many members of Congress, it also created some tension. In essence, by striving to serve all members of Congress, OTA served no member in particular. Thus, as we will discuss in a later section, during difficult times only a few members were willing to take political risks to support OTA.

Finally, although the vast majority of OTA's reports were well received, a relatively small fraction of reports proved to be highly controversial. Almost all these studies focused on the role of technology in national security issues. In particular, OTA published a series of reports during the 1980s that were highly critical of President Reagan's proposed Strategic Defense Initiative (OTA 1985a, 1985b, 1988). These reports laid a political minefield for OTA that would later be an important factor in its demise. As Gibbons recently recalled during an interview on National Public Radio, the set of SDI reports "was the most vexing and controversial thing I think we did in the 14 years that I was there" (National Public Radio 2001).

In 1993 Gibbons departed OTA to become President Clinton's White House Science Advisor. At the time OTA appeared to be firmly established on the Hill. Gibbons's successor was Roger Herdman, a former vice president of the Sloan-Kettering Cancer Center and associate director under Gibbons. Herdman continued in Gibbons's footsteps, and he began to implement some changes at OTA in response to OTA's self-study. He also established a Long Range Planning Task Force to examine alternative structures for OTA's staff, eventually flattening out OTA's organization even further, based on the task force's recommendations (Wood 1997).

Under Herdman, budgetary pressures within Congress resulted in shrinking budgets at OTA and consolidation within the agency that left little fat. Between 1993 and 1995, OTA reduced its middle and senior management ranks by 40%. At the operating level, studies that once could afford to devote as much as half their budget to field visits and small external contracts for supporting analyses were, by 1995, devoting approximately 80% of their budget to staff salaries. At the time it was closed, OTA had, for several years, been living off the accumulated intellectual capital of its staff. It is unlikely that it would have been able to sustain its level of output and quality without an increase in budget.

Accomplishments of OTA

Over its 23-year history OTA had a number of accomplishments. In terms of written work, OTA published almost 750 full assessments, background papers, technical memoranda, case studies, and workshop proceedings. Full assessments, which were comprehensive analyses of book length, were the most visible product. From start to finish, a full assessment consumed some 18–24 months and cost approximately $500,000 in direct costs. In general its reports were highly regarded. For example, between 1992 and 1994, 12 OTA assessments won the National Association for Government Communicators' prestigious Blue Pencil Award (Houghton 1995). OTA's accomplishments, however, go far beyond its written works. In particular, it (1) raised the level of debate in Congress and influenced the legislative process through testimony and other less formal means of communication, (2) saved U.S. taxpayers many times its cost, and (3) served democracy by bringing a wide range of views into many debates on science and technology issues. Here we look at these three areas in more depth.

Effect on the Legislative Process

Quantifying the effect of any single actor or report on the legislative process is difficult. In particular, the types of analysis that OTA conducted would often be factored into the process in subtle ways. Typically, congressional staff, as opposed to members of Congress, would be the ones with enough time to become familiar with the details of specific OTA reports. In addition, throughout OTA's history, personal contacts, telephone calls, meetings, briefings, and testimony typically supplemented physical reports. Largely through these types of activities, OTA raised the level of debate in Congress on science and technology issues. Although the effects of the range of activities carried out by OTA are difficult to quantify, a number of studies have tried to do so. For example, two studies have been carried out involving congressional staff: Congress's 1978 internal review of OTA and Bruce Bimber's 1990 survey of attitudes about the work of OTA. In addition, in 1995 OTA attempted to look at the legislative effects of its assessments carried out during the early 1990s (OTA 1995b).

In 1978 Congress carried out an internal review of OTA. OTA's relationship with Congress was shaky at this time. This review involved surveying 47 congressional staff members to determine how useful the 37 OTA reports published between 1972 and 1977 had been in assisting Congress (U.S. Congress 1978). Only staff members who had used one or more OTA assessments were included in the survey. As shown in Figure 3-2, more than half of the staff members in the survey rated the OTA assessment reports as "very useful" and another 25% rated them as "useful". In addition, 50–80% of the staff members were satisfied with the effect, content, working relationship, and timing of the reports (U.S. Congress 1978).

In 1990 Bruce Bimber conducted a study on OTA for the Carnegie Commission on Science, Technology, and Government (Bimber 1990). At this time, OTA was well established within Congress. This study involved surveying 35 congressional staff members. All of the staff included in the study were regularly involved in science and technology issues. The survey evaluated staff views and attitudes about the work of all four legislative support agencies: the Congressional Budget Office, the Congressional Research Service, the Government Accounting Office, and OTA. Here we will focus on its results related to OTA.

As shown in Figure 3-3, half of the staff members in the survey found OTA reports to be "very useful," and another 41% rated them as being "useful." In addition, 80–100% of the staff members felt that OTA reports provided a unique contribution, were technically competent, maintained political neu-

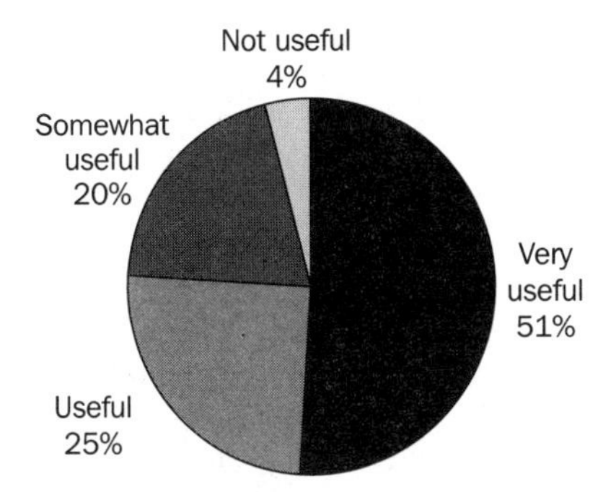

Figure 3-2. Congressional Staff Ratings of 37 OTA Reports (1972–1977)

Source: Adapted from U.S. Congress 1978, 126–127.

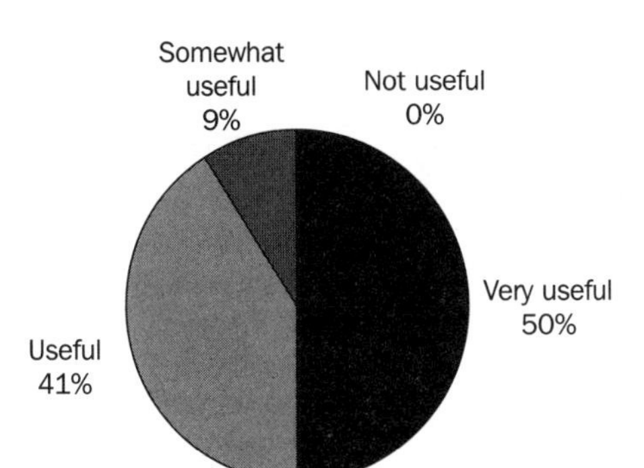

Figure 3-3. Congressional Staff Ratings of OTA Reports

Source: Adapted from Bimber (1990).

trality, and were available in time to be useful (Bimber 1990). Comparing the results of the 1978 review with Bimber's 1990 study suggests that between the late 1970s and early 1990s OTA's reputation among congressional staff, who deal with science and technology issues on a regular basis, improved considerably.

In 1995, OTA reviewed the legislative effect of 76 of its reports produced during the early 1990s (OTA 1995b). This review was conducted by OTA for use in its ensuing budget battles. In particular, the review focused on three areas where OTA reports have had significant effects: (1) as a source of background material in drafting specific pieces of legislation, (2) as the basis of testimony used in hearings, and (3) where reports were cited or praised by members of Congress on both sides of the aisle. Some examples of typical effects are shown in Table 3-2.

Finally, OTA's effect on the legislative process went beyond its reports. In particular OTA staff often testified during congressional hearings and engaged in other less formal modes of communication with members of Congress and their staff. For example, during FY1994, OTA delivered 51 publications to

Table 3-2. Selected OTA Reports and Their Legislative Effects

Report title	*Legislative effects*
Redesigning Defense: Planning the Transition to Future U.S. Defense Industrial Base	Helped form the basis of the federal Acquisition Streamlining Act of 1994, used in hearings before Senate Subcommittee on Defense Industry and Technology
After the Cold War: Living with Lower Defense Spending	Used in a bipartisan way to help shape many parts of the defense conversion legislation in the Defense Authorization and Appropriation Acts of 1993–1995, used as a basis for testimony a number of times
Industry, Technology and the Environment: Competitive Challenges and Business Opportunities	Cited as influencing three pieces of legislation: H.R. 3813, S. 2093, and S. 978
Making Government Work: Electronic Delivery of Federal Services	The Food Stamp Fraud Reduction Act of 1993 referred explicitly to this report. The report was the subject of two hearings in the Senate
Research Health Risks	Used to develop language in the Risk Assessment Improvement Act and the Risk Communication Act of the 103rd Congress
Advanced Liquid Metal Reactor	Cited by the FY1995 Energy and Water Appropriations Bill to discontinue funding for the ALMR
Difficult-To-Reuse Needles for the Prevention of HIV Infection among Injecting Drug Users	Led to the requesting subcommittee's decision to abandon proposed legislation

Source: Adapted from OTA 1995b.

Congress, and OTA staff testified 38 times before congressional committees (Herdman 1995). This was typical of OTA. During the 1980s and early 1990s, OTA staff testified between 35 and 55 times per year (Herdman 1995; Nichols 1990). In addition, OTA staff provided special briefings and expert advice to Congress on a regular basis. As Bimber (1990, 31) points out, almost 80% of the congressional staff members he surveyed considered person-to-person communication with OTA staff to be "just as important" as or "more important" than written reports.

Saving Money

From early on, OTA's supporters argued that by improving the decisionmaking process, ultimately OTA would save U.S. taxpayers many times more that its cost. For example, when testifying before the Senate Committee on Rules and Administration in 1972, Senator Gordon Allot (Republican, Colorado) argued that

> ... the Office of Technology Assessment will not only enable us to spend wisely on behalf of science, but convince the public whose money we are spending that our spending is done intelligently and conscientiously. The money it will cost to establish and operate the Office of Technology Assessment should be recouped many times over in expenditure savings. (Leg. Hist. P.L. 92–484, 3578)

The issue of cost saving was a recurring theme throughout the history of OTA. Yet, because OTA reports often influenced the process in subtle ways, it has been difficult to quantify the actual cost savings produced by OTA. In fact, it was not until OTA was threatened in 1995 that it tried to quantify some of the cost savings that it had helped to bring about. Whereas it is not possible to say precisely how much money OTA saved U.S. taxpayers, it is possible to point out examples where OTA studies were instrumental in clarifying issues that helped Congress make cost-saving decisions. A few examples of studies contributing to significant cost savings are the following (OTA 1995a, 1995b):

- OTA's study "Special Care Units for People with Alzheimer's and Other Dementias" helped state and federal governments realize annual cost savings of $14 million.
- OTA's study "The Social Security Administration's Decentralized Computer Strategy: Issues and Options" helped Congress to evaluate the Social Security Administration's massive computer procurement strategy, leading to a total savings of $368 million. The House Committee on Appropriations report cited the guidance provided by this OTA report.
- OTA's study "After the Cold War: Living with Lower Defense Spending" recommended technical changes on base closures that helped Congress save tens of millions of dollars. This report was used by the Republican and Democratic Defense Conversion Task Forces alike, and it helped to shape many parts of the defense conversion legislation in the Defense Authorization and Appropriations Acts of 1993, 1994, and 1995.
- OTA conducted a series of studies during the early 1980s that raised caution about the Synthetic Fuels Corporation and helped secure approximately $60 billion in savings.

These examples barely begin to scratch the surface of the cost savings OTA helped contribute by improving the decisionmaking process. Even if only a small fraction of these savings were credited to OTA, one could easily argue that OTA more than paid for itself during its existence.

Serving Democracy

When OTA's assessments were effectively integrated into the political process, they functioned as a vehicle of democratic decisionmaking rather than of technocratic planning by experts.[6] In essence, OTA assessments provided a relatively open means of publicly scrutinizing experts, encouraging greater care in documenting underlying assumptions and an even-handedness in dealing with conflicting views. For example in a typical year OTA brought approximately 5,000 people into its process. These experts provided advice, took part in workshops, served as members of advisory panels, and participated in OTA's peer review process. They came from all walks of life: academia, industry, public interest and citizen groups, and government. Yet OTA did not use expert advice to come up with a set of policy recommendations. Instead, it tried to provide policymakers with a well-reasoned set of options that fairly represented the full range of opinions about a given issue. Meanwhile, the final responsibility for making policy choices remained in the hands of elected representatives.

On the surface it may seem counterintuitive that institutionalizing technology assessments through OTA would make the legislative process more democratic. However, OTA did just that, largely by helping Congress overcome the difficulties of incorporating expertise into the policymaking process. Although OTA helped to make decisionmaking more rational, it did not substitute a purely technical approach for the political process. One could argue that OTA was an attempt by Congress to balance the technocratic and democratic tendencies within its ranks (Andrews 2002).

The tension between technocratic and democratic approaches to decisionmaking is a common phenomenon that has been dealt with to varying degrees in the different branches of government.[7] Those who believe in the democratic ideal argue that the average citizen can be sufficiently informed on technical issues to play as important a role as experts in the policy process (for example, see Sclove 1995). Alternatively, those who believe in the technocratic ideal would argue that the solution to poor decisionmaking in government on scientific and technical issues is to get more and better science into the decisionmaking process. The idea of technology assessment was rooted in the technocratic ideal, but the realization of technology assessment in OTA was somewhere between these two poles.

Shifting Power in Congress Leads to the Demise of OTA

The 1990s brought about a dramatic shift of power in Congress. This shift, which took place in the context of a rising current of conservative thought and ideology and general dissatisfaction with the first two years of the Clinton administration, culminated in the 1994 congressional elections and led directly to the demise of OTA. From early on, the newly empowered Republican leadership in Congress announced their intention to eliminate OTA. Abolishing OTA was part of the new leadership's broader aim, as outlined in its *Contract with America*, of downsizing government (Gingrich 1994). When Congress began holding hearings during early 1995 on downsizing the legislative branch support agencies, conservative thinkers from the Heritage Foundation, American Enterprise Institute, and Citizens Against Government Waste figured prominently (U.S. Congress 1995). During these hearings three main arguments were used against OTA. Proponents of eliminating the agency argued that (1) Congress should eliminate OTA to demonstrate a willingness to clean up its own house before looking elsewhere for more draconian cuts, (2) OTA reports took too long to complete, and (3) OTA merely duplicated work done elsewhere, both inside and outside of government.

Cleaning Up Its Own House

During 1995, Congress focused primarily on short-term fiscal constraints, i.e., balancing the federal budget. Although all sides agreed that meeting this goal would require considerable reductions throughout the federal system, there was considerable disagreement about precisely where to make cuts. Early on it became clear that Congress would need to look in its own backyard before looking elsewhere. As Senator Connie Mack (Republican, Florida), a vocal critic of OTA, put it,

> If Congress is going to make the necessary reductions to the entire Federal Government and return to fiscal sanity, if Congress is going to mandate that Executive departments and agencies become more focused and efficient, and if Congress is going to return to the American people control of their lives, then Congress must demonstrate the political will and leadership by putting its own house in order. (U.S. Congress 1995)

Thus, eliminating OTA took on a symbolic value. OTA had a relatively small budget—about $20 million—but its small size and highly specialized focus helped make it a relatively defenseless target. By eliminating OTA, Congress could demonstrate that it would not exempt itself from frugality.

Timing of Reports

As discussed above, OTA often faced criticism that its studies took too long to complete and thus did not meet the immediate needs of members of Congress. During the mid-1990s budget battles over OTA, this issue reemerged. For example, in December 1994, Representative Robert Walker (Republican, Pennsylvania), the newly appointed Chair of the House Science Committee, stated that for OTA to survive, its studies would have to become more short-term and responsive to the appropriations process (press conference, December 14, 1994). However, one of OTA's distinguishing features was that it attempted to take a forward-looking, long-term perspective on science and technology issues. To meet this goal, OTA's studies required a considerable amount of time to complete, typically 1–2 years. As Representative George Brown (Democrat, California), a long-time member of the TAB, observed,

> OTA was established to provide comprehensive, balanced analysis of complex questions. It looked at the technology, at its social and economic impacts, and then made a range of recommendations for Congressional action. That process takes a long time. (*Congressional Record* 1995)

In a sense, OTA was caught in a no-win situation. To survive, it would need to focus on conducting short-term studies. In fact, during the early 1990s, in response to congressional pressures, OTA began to shift toward conducting short-term studies; however, shifting further in this direction would make it more difficult for OTA to distinguish itself from other information sources such as the Congressional Research Service. In turn, this solution would feed into claims by critics of OTA that it simply duplicated work done elsewhere.

Duplication of Work

Finally, during the debates on downsizing the legislative branch support agencies, critics of OTA argued that it simply duplicated work done elsewhere (U.S. Congress 1995). The critics noted that before OTA was created Congress already had to deal with science and technology issues. However, if they did not have internal expertise in a particular area (in committee staff or one of the existing congressional support agencies) then they relied on executive branch agencies such as the National Science Foundation (NSF) or other organizations such as the National Academy of Sciences (NAS).

Yet, as discussed above, during the late 1960s and early 1970s Congress began to realize that the existing sources of information on science and technology issues were insufficient. At the time, extensive discussions ensued within Congress about expanding the capacity of the Congressional Research Service (CRS), the General Accounting Office, or both to better handle complex questions about science and technology. However, given the inherent forward-looking and long-term perspective required for the technology assessment process, neither of these organizations was deemed appropriate for taking on the tasks. During the mid-1990s, when the OTA was shut down, the alternatives to OTA that were available would have required significant modifications to take on the functions of OTA. The main features that distinguished these other options from OTA include the following:

- The General Accounting Office (GAO) is an organization of auditors and accountants who focus primarily on management of government programs and waste, fraud, or abuse. Although GAO has some technical expertise, it has little policy–analytic expertise.
- The Congressional Research Service (CRS) focuses on responding to requests from members of Congress, usually with quick turnaround time. Typically their reports are much shorter and provide less in-depth analysis than OTA reports.
- The Congressional Budget Office (CBO) provides budget data and analyzes alternative fiscal and budgetary effects of legislation. It is staffed almost entirely by economists and accountants.
- The President's Office of Science and Technology Policy (OSTP) does not serve congressional needs. Instead it supports the policies and views of the incumbent president and his or her advisers.

- The National Science Foundation (NSF) carried out technology assessments during the 1970s under its Research Applied to National Needs (RANN) Program and during the latter years of the Nixon administration after the White House Science Office and Presidential Science Advisory Committee were abolished. It never developed significant analytical staff, and RANN was constantly in the position of intruding on the turf of other agencies, which in turn left it vulnerable to considerable criticism (Coates 1982).
- The National Academy of Sciences (NAS) is perhaps the organization most similar to OTA. However, there are a number of key differences: (1) NAS reports are typically written by a panel of experts who may depend on the agency backing the study; (2) NAS panels are usually made up of individuals whose views are near the center of the spectrum on a given topic to ensure that a consensus can be reached (in contrast, OTA strove to represent the range of views among experts); (3) NAS panels usually determine the scope of their work, the methods they use, and the questions they choose to answer in response to a given request; (4) NAS studies often make specific recommendations as opposed to laying out and evaluating a range of policy options; and (5) the NAS has a separate constituency in the elite scientific community, which has different and independent interests from those of Congress and the public.

Subsequent chapters of this book will explore in detail how these (and other) alternatives might be modified to fill the void left by the closure of OTA. When OTA was closed, though, Congress did not shore up any of the alternatives to fill this void.

Conclusion

What has been lost with the demise of OTA? First and foremost, OTA provided a relatively unbiased and responsive force within Congress, aimed at raising the level of debate on complex technical issues. Second, OTA served Congress and indirectly the citizens of the United States by making the legislative process more democratic, that is, inclusive of a wide range of interested parties. Third, OTA helped to counteract the previously existing imbalance of power on technical issues between the executive and legislative branches of government.

The need for Congress to be able to evaluate science and technology issues in a forward-looking manner, independent from executive agencies and special interest groups, is as real today as it was 25 years ago. The rationale for creating OTA in 1972 still rings true today:

> The Congress has difficulty in comprehensively envisioning all of the potential influences of technology; the subject has many facets and requires for its understanding long study in many different and special-

ized technical and professional disciplines. Not only are the members of Congress seldom trained to ascertain the optimum use of technology, but they are not adequately served by a mechanism independent of the Executive Branch or special interest groups to help them make such judgments. (Leg. Hist. P.L. 92–484, 3573)

Clearly, the potential sources of information on science and technology issues—from the media, the executive branch, and special interest groups—have expanded considerably during the past few decades, but one could argue that it is more difficult today than it was two and a half decades ago to sort through this sea of information.

Despite OTA's contributions, its relatively small size and highly specialized tasks made it an easy target for congressional budget cutters. It is important to realize the strong relationship between a particular moment in the agenda of the Republicans who gained the majority in Congress during the 1994 midterm elections and perceived deficiencies in OTA's work.[8] Understanding the forces that led to the demise of OTA can help clarify what, if anything, should replace it. If OTA suffered from fundamental flaws, then a new office could arise out of an improved structure. If OTA suffered from a unique confluence of events, then a similar structure could be reconstituted. If the entire congressional environment changed, then perhaps no similar institution could succeed.

Notes

[1]A number of good sources provide insight on the creation of OTA. See Bimber (1996); Coates (1982); Gibbons and Gwin (1988); Kasper (1972); Kunkle (1995); and U.S. Congress (1972, 1978).

[2]Congress also established the Select Committee on Government Research in 1963, chaired by Representative Carl Elliott (Democrat, Alabama), which recommended establishing a standing Joint Committee on Research Policy as a counterweight to the White House Office of Science and Technology Policy, as well as a new science subcommittee in the House Armed Services Committee and the Science Policy Research Division in the Legislative Reference Service (the forerunner to the Congressional Research Service). Daddario's subcommittee became the launching pad for significant reforms, including requests of the Committee on Science and Public Policy (COSPUP) at the National Academy of Sciences to study the relationship between basic research and national goals, and the National Academy of Science's report, *Technology: Process of Assessment and Choice* (1969), which helped lead to the creation of OTA (Bimber 1996; Guston 2000).

[3]During the 1970s the staff of the Congressional Research Service (CRS) and the General Accounting Office (GAO) were expanded considerably. In addition, Congress created the Congressional Budget Office (CBO) in 1974.

[4]See "The Debate over Assessing Technology," *Business Week* (April 8, 1972), available in OTA (1995c).

[5]See Wood (1997) for specific evaluations of each stage of the process.

[6]There are, however, criticisms of OTA's interest group, rather than participatory, approach to technology assessment (e.g., Bereano 1997).

[7]For discussions of how this tension has been dealt with in the executive branch, see Jasanoff (1995), and in the legislative branch, see Casper (1976) and O'Brien (1982).

[8]A survey conducted by Andrews (2002) supports the importance of this partisan moment.

References

Andrews, Clinton J. 2002. *Humble Analysis: The Practice of Joint Fact-Finding.* Westport, CT: Praeger.

Bereano, Philip L. 1997. Reflections of a Participant–Observer: The Technocratic–Democratic Contradiction in the Practice of Technology Assessment. *Technological Forecasting & Social Change* 54: 163–176.

Bimber, Bruce. 1990. Congressional Support Agency Products and Services for Science and Technology Issues: A Survey of Congressional Staff Attitudes about the Work of CBO, CRS, GAO, and OTA. Paper prepared for the Carnegie Commission on Science, Technology, and Government, New York.

———. 1996. *The Politics of Expertise in Congress: The Rise and Fall of the Office of Technology Assessment.* Albany: State University of New York Press.

Casper, Barry. 1976. Technology Policy and Democracy. *Science* 194: 29–35.

Coates, Vary. 1982. Technology Assessment in the National Government. In *The Politics of Technology Assessment*, edited by David O'Brien and Donald Marchand. Lexington, MA: Lexington Books.

———. 1999. Technology Forecasting and Assessment in the United States: Statistics and Prospects. *Futures Research Quarterly* 15: 5–25.

Congressional Record. 1995. September 29.

Gibbons, John, and Holly Gwin. 1988. Technology and Governance: The Development of the Office of Technology Assessment. In *Technology and Politics*, edited by Michael Kraft and Norman Vig. Durham, NC: Duke University Press.

Gingrich, Newt. 1994. *Contract with America.* New York: Times Books.

Guston, David H. 2000. *Between Politics and Science: Assuring the Integrity and Productivity of Research.* New York: Cambridge University Press.

Herdman, Roger. 1995. Testimony before the U.S. Congress Senate Appropriations Committee, May 26.

Herdman, Roger C., and James E. Jensen. 1997. The OTA Story: The Agency Perspective. *Technological Forecasting & Social Change* 54: 131–144.

Houghton, Amo. 1995. *Congressional Record.* September 28, pp. E1868–E1870.

Jasanoff, Sheila. 1995. *The Fifth Branch: Science Advisers as Policymakers.* Cambridge, MA: Harvard University Press.

Kasper, Raphael (ed.). 1972. *Technology Assessment: Understanding the Social Consequences of Technological Application.* New York: Praeger.

Kunkle, Gregory. 1995. New Challenge or the Past Revisited? The Office of Technology Assessment in Historical Context. *Technology in Society* 17(2): 175–196.

National Academy of Sciences. 1969. *Technology: Process of Assessment and Choice.* Washington, DC: National Academy of Sciences.

National Public Radio. 2001. All Things Considered. Transcript of the broadcast of July 18, Burrelle's Information Services, Box 7, Livingston, NJ 07039.

National Science Foundation (NSF). 1993. *Federal Funds for Research and Development: Detailed Historical Tables—Fiscal Years 1956–1993.* Washington, DC: National Science Foundation.

———. 2000. *National Pattern of Research and Development Resources: 2000 Data Update.* Report No. NSF 01–309. Washington, DC: National Science Foundation.

Nichols, Rodney. 1990. *Vital Signs OK: On the Future Directions of the Office of Technology Assessment.* Paper prepared for the Carnegie Commission on Science, Technology, and Government, New York.

Norman, Colin. 1977. O.T.A. Caught in Partisan Crossfire. *Technology Review* October/November.

O'Brien, David. 1982. The Courts, Technology Assessment, and Science–Policy Disputes. In *The Politics of Technology Assessment*, edited by David O'Brien and Donald Marchand. Lexington, MA: Lexington Books.

Office of Technology Assessment (OTA). 1977. *Technology Assessment in Business and Government: Summary and Analysis.* PB–273164. Washington, DC: U.S. Government Printing Office.

———. 1979. *OTA Priorities, 1979.* Washington, DC: U.S. Government Printing Office.

———. 1985a. *Ballistic Missile Defense Technologies.* Washington, DC: U.S. Government Printing Office.

———. 1985b. *Anti-Satellite Weapons, Countermeasures, and Arms Control.* Washington, DC: U.S. Government Printing Office.

———. 1988. *SDI: Technology, Survivability, and Software.* Washington, DC: U.S. Government Printing Office.

———. 1993. *Policy Analysis at OTA: A Staff Assessment.* Washington, DC: Office of Technology Assessment.

———. 1995a. *About OTA.* Washington, DC: Office of Technology Assessment, unpublished.

———. 1995b. *Legislative Impact Summary.* Washington, DC: Office of Technology Assessment, unpublished.

———. 1995c. *OTA Legacy* (a five-volume CD collection). 052–003–01457–2. Washington, DC: U.S. Government Printing Office.

Sclove, Richard E. 1995. *Democracy and Technology.* New York: Guilford Press.

U.S. Congress, House of Representatives Committee on Appropriations. 1995. *Downsizing Government and Setting Priorities of Federal Programs.* Washington, DC: U.S. Government Printing Office.

U.S. Congress, House of Representatives Committee on Science and Technology. 1978. *Review of the Office of Technology Assessment and Its Organic Act.* Washington, DC: U.S. Government Printing Office.

U.S. Congress, Senate Committee on Rules and Administration. 1972. *Technology Assessment for the Congress.* Washington, DC: U.S. Government Printing Office.

van Dam, Laura, and Robert Howard. 1988. How John Gibbons Runs through Political Minefields: Life at the OTA. *Technology Review* October: 47–51.

von Hippel, Frank, and Joel Primack. 1991. Scientist as Citizen. In *Citizen Scientist*, edited by Frank von Hippel. New York: Simon & Schuster.

Weingarten, Fred. 1995. Obituary for an Agency. *Communications of the ACM* 38(9): 29–32.

Wood, Fred B. 1997. Lessons in Technology Assessment: Methodology and Management at OTA. *Technological Forecasting & Social Change* 54: 145–162.

4

Insights from the Office of Technology Assessment and Other Assessment Experiences

David H. Guston

Creating or recreating a scientific and technical advisory apparatus for Congress involves the specific nature of scientific and technical advice and the relationship between providing this advice and the design of institutional relationships to do so. Chapter 2, by Smith and Stine, covers the surprisingly long sweep of the history of science advice to the U.S. Congress, describing actions that Congress has taken to ensure that it has access to technical expertise. Subsequent chapters articulate a suite of possible alternatives for the new institutionalization of such a capacity, ranging from a new and improved Office of Technology Assessment (OTA) to a system of distributed analysts performing for Congress.

This chapter attempts to fill the considerable gap between the history of congressional action on scientific and technical advice and proposals for new institutions. We should learn something from the critical appraisal of experience to help with planning new enterprises. The argument is based on two principles. First, as Bruce Bimber (1996) writes in his analytical history of OTA's life cycle, "The degree of politicization of expertise may be more an institutional phenomenon than a product of the preferences or style of politicians, the moral or professional commitment of experts, or an inexorable trend away from neutrality." Second, liberal-democratic governance is under some obligation to be informed about the causes and effects of its own operation, and this obligation extends to understanding the intellectual underpinnings of public action.[1] I believe that the enterprise of the June 2001 workshop that inspired this book mobilizes these principles.

This chapter begins with a synthesis of some relevant scholarship about the institutionalization of expertise and the conduct of technically sophisticated policy analysis and assessment that, written mostly after OTA's demise, may be relevant for the recreation of a congressional science advisory capacity. Based on this discussion and the histories in Chapters 2 and 3, it then identifies and explores the issues that proposals for new institutions for advice, analysis, and assessment will likely need to address.

Institutionalizing Technology Assessment

The demise of OTA provides the opportunity to reconsider the provision of scientific and technical advice, analysis, and assessment to Congress, both organizationally and conceptually. Other chapters in this volume discuss new organizational options directly (also see Hill 1997; La Porte 1997). Conceptually, the years since OTA's demise provide the opportunity to inform consideration of new organizations with more recent scholarship and other developments in technology assessment. Many of these developments have taken place in the agendas and performance of technology assessment organizations in other countries, which Vig (Chapter 5 of this volume) and Vig and Paschen (2000) deal with directly. Here I discuss four of these developments: public participation in technology assessment; new styles of technology assessment, including constructive and real-time technology assessment; the practice of assessments for both political and technical virtuosity; and, similarly, the structuring of institutions to produce politically and technically virtuous assessments.

Public Participation in Technology Assessment

The first important development concerns the increasing role of lay citizens in the process of assessment or analysis. Although OTA made extensive use of stakeholders as members of panels and reviewers of drafts, it made little effort to include lay citizens in its work (Bereano 1997). Nevertheless, other technology assessment practitioners have adopted—often to good effect—participatory methods such as citizens' panels (Guston 1999; Hörning 1999; Joss and Durant 1995), scenario workshops (Andersen and Jaeger 1999; Sclove 1999), and focus groups (Dürrenberger et al. 1999). In aggregate, these participatory methods are also known as interactive technology assessment (Grin et al. 1997).

How the policy–analytic and public deliberation versions of technology assessment accommodate one another is an important, but open, question intellectually and practically. Vig and Paschen (2000) refer to these two styles as the "instrumental" and "deliberative" modes of technology assessment, respectively. There is, however, no necessary competition between the two models and, moreover, there ought to be complementarities (Guston and Bimber 1998). Whereas participatory mechanisms offer little chance of serving as more than brokers of analysis that has been performed by more expert actors, they do offer the prospect of creating broad, novel frames and insight into public attitudes about the acceptance of or hostility toward new technologies. It seems likely, then, that "public policy is best served by the flourishing of both enterprises," but there need be no presumption "that both enterprises must fit comfortably in the same institution" (Guston and Bimber 1998).

Planning for a new capacity for congressional scientific and technical advice and analysis should confront this challenge of finding ways in which the participatory and analytic modes complement each other. It is plausible that, as in some of the European experiences, participatory mechanisms are important for public education around a scientific and technical issue, and such an educative role may also contribute to building a broader constituency for analysis. However, in OTA's experience, effective, successful stakeholder participation was time-consuming and expensive, and critics found OTA's work neither timely nor cheap. It is unclear whether broader participatory mechanisms outperform stakeholder participation on these criteria and to what extent they could and should be incorporated into a new technology assessment institution serving Congress.

New Styles of Technology Assessment

A second important development is about increasing the interaction between assessment and analysis on one hand and the design of new technologies on the other. With enough forethought and lead time, interactive modes of technology assessment, coupled with expert modes, can serve a constructive role in technological and societal choice—maximizing the benefits and minimizing the problems that may be associated with knowledge-based innovation. This "constructive technology assessment" (Schot and Rip 1997) does not conceive of technologies as preformed black boxes to which society must adapt, but rather as more flexible entities that are coproduced by the social contexts of their invention and use. It attempts "to broaden the design of new technologies" through "[f]eedback of TA [technology assessment] activities into the actual construction of technology" (Schot and Rip 1997). The tenets of constructive technology assessment include socio-technical mapping, a combination of traditional stakeholder analysis with the plotting of technical activities; early and controlled experimentation to identify unanticipated consequences and, if need be, ameliorate them; and interaction between innovators and the public (as described above) to articulate better the demand side of technology development.

More recently, Guston and Sarewitz (2001) have continued on this trajectory to describe "real-time technology assessment," which conducts historical and social scientific research in direct collaboration with the natural science and engineering work being assessed. Real-time technology assessment differs in three ways from constructive technology assessment: (1) although it engages in socio-technical mapping and demand-side articulation, it does not involve experimentation because its focus is the knowledge-creation process itself; (2) it uses a variety of social scientific methods to investigate how public knowledge, perceptions, and values about emerging technologies change over time; and (3) it integrates retrospective, historical work on the social impact of innovation with prospective scenario analysis to render contemporary innovation more amenable to understanding and modification.

Clearly, a congressional advisory mechanism of whatever makeup should not be directly involved in constructive or real-time technology assessment. That is, the staff of such an organization should not themselves collaborate with natural scientists and engineers for the purpose of steering the latter's research. However, if the prospective angle envisioned by OTA's legislative charter is to be retained at all, the new mechanism may find constructive and real-time technology assessment interesting and appropriate methods with which to experiment, and it may consider policies that encourage such activity elsewhere. Moreover, the collaboration between social science and natural science in the service of technology assessment and public information is an ideal to which any institutionalized technology assessment should aspire.

The Practice of Assessments

A third development, alluded to by Smith and Stine (2003), concerns how assessments or analyses may be conducted to achieve both political and technical goals. Until recently, the literature on technology assessment and policy analysis has neglected the relationship among intellectual function, analytical process, and institutional form. Objectivity, roughly synonymous with Smith and Stine's "disinterestedness," was seen as either an intellectual standpoint or, if associated with process or structure at all, was attributed to distance or insulation from interested parties. Such insulation would, however, render the demands of agenda setting and relevance almost insuperable.

This situation creates what Guston and Bimber (1998) refer to as "the dilemma of expert independence": the demand by the policymaking consumers of analysis to maintain control over the agenda of experts and over the process of interest aggregation and representation; and the countervailing demand by the producers to be independent in their production of the analysis. The dilemma must be addressed, however, because of the mutual interest of the policymakers and experts, not to mention citizens, in relevant analysis for decisionmaking.

A variety of scholarship over the past decade has begun to identify procedural and institutional factors that promote disinterested analysis while still satisfying requirements of relevance as well. Projects organized by William C. Clark on "social learning and the environment" and on "global environmental assessments" (GEA) have been at the forefront of such scholarship (although they focus exclusively on international environmental assessments, which overlap at least somewhat but not necessarily completely with technology assessments).[2] The Social Learning Group (2001) taps a broad array of national case studies in both the developed and the developing world to examine the interplay of ideas, interests, and institutions in the practice of environmental management. Among their conclusions is that many factors beyond the technical adequacy of environmental assessments, that is, the capacity of local institutions to learn from assessments, move nations to informed environmental action.

The GEA project has focused on the design and management of effective assessments and the information systems that link global environmental assessments to local decisionmaking (GEA 2000). It will provide a variety of case studies and commentary aimed at improving the practice of environmental assessment (Farrell and Jaeger 2003). GEA research suggests that "much about what makes some assessments more effective than others seems to be tied up with the *process* by which they are developed, rather than just the *product* itself" (Clark and Dickson 1999; emphasis in the original). GEA research also defines criteria for good assessments—saliency, credibility, and legitimacy—that it finds to be products of the procedural elements of an assessment, including when in the evolution of an issue an assessment is conducted, how an assessment structures its audience, and how an assessment manages to negotiate the interface between politics and science (Clark and Dickson 1999; Cash et al. 2003).

Approaching the same topic of science and technology analysis from a different direction, engineer and planner Clinton Andrews (2002) argues that the route to successful analysis is the "practice of joint fact-finding." By studying cases in comparative risk assessment and electric power regulation in addition to OTA, Andrews concludes that enabling analysts and stakeholders to collaborate leads to the production of adequate, valuable, effective, and, most importantly, legitimate analysis. Despite OTA's solution to the analyst's technical problem of modeling reality in a method acceptable to a variety of clients, Andrews also points out that OTA did not completely solve the analyst's communication problem when it did not activate a broad enough array of latent congressional clients.

This emphasis on social learning, communication, and the process of assessment leads to different ways of evaluating assessments. It displaces attention from the bound volume of the report to the greater variety and forms of communication, including the interactions that produced the report in the first place. Thus, when critics point to the "useless" and tardy book-length OTA report that failed to change a congressional vote or alter a program budget, they adopt a discredited "silver bullet" account of policy analysis. A full evaluation of policy analysis or technology assessment includes not only these "actual impacts" of the study, but also its more nuanced impact on general thinking about the issue (e.g., how an issue is framed), as well as the learning engaged in by participants in the process (including analysts and stakeholders alike) and nonparticipants (the targets of the advice as well as the general public).[3] With OTA, it was often felt that the report was important significantly in that it represented a great deal of negotiation and learning among analysts, staff, and stakeholders that increased knowledge and reduced conflict in preparation for congressional action.

The Structuring of Institutions

In a related way, other recent scholarship has addressed how institutions can be structured to promote the effective production and use of relevant and disinterested advice and analysis. Sheila Jasanoff (1990), for example, in her account of science advice in executive agencies, anchors the helpfulness (if not the objectivity) of advisory committees in accounts of successful "boundary work" (see also Gieryn 1999). In the context of advising such agencies as the U.S. Environmental Protection Agency (EPA) and the U.S. Food and Drug Administration, successful boundary work generally means the parsing of the distinction between science and policy—and thus the respective roles of science advisers and policy decisionmakers—in a more, rather than less, ambiguous way. That is, science advisers are more successful when they and their staff suppress rather than make real any distinction between science and policy. Jasanoff also finds that, as in OTA's experience, the social aspects of peer review, balancing interests, and stakeholder participation contribute to the technical and political credibility of science advice.[4]

Jasanoff (1990) points to another organization intimately involved in negotiating the complexities of regulatory science, the Health Effects Institute (HEI). Jointly funded by EPA and the automobile industry and bolstered by prominent and interdisciplinary advisory panels that provide peer review, HEI has established itself as a credible sponsor and broker of research relevant to regulatory decisions that incorporate the health effects of air pollution. Following a review by the National Research Council (1993), HEI continued to improve its credibility by broadening, rather than narrowing, its engagement with stakeholders and its efforts in producing relevant, timely research (Keating 2001).

Such attention to the design of institutions providing scientific and technical advice for executive functions is vital, especially if one considers Bimber's (1996) argument about the natural trend of executive agencies to move toward the politicization of expertise. OTA demonstrated that Congress, on the other hand, could encourage a trend toward neutral expertise by forcing responsiveness to the diverse ideological and jurisdictional agendas of two parties and multiple committee chairs. In Bimber's argument, this studied structural neutrality, both manifest in and managed by OTA's Technology Assessment Board (TAB), still did not permit OTA to provide the highly particularized informational products that might have extended or ensured its existence. Similarly, Andrews (2002) concludes that OTA solved the analyst's technical problem but not his or her communication problem. OTA was structurally and intellectually neutral, but it may have provided too generalized a benefit for a particularized institution such as Congress.

In addition to scientific advisory committees, HEI, and OTA, a variety of other organizations exist "between politics and science." In other work (Guston 2001), I attempt to formulate a more general theory of such boundary organizations that (1) exist at the mutual frontier of politics and science but have strict lines of accountability into each; (2) involve the participation of actors from both sides in addition to professionals who serve a mediating role; and (3) produce goods and services of value to actors on both sides. Drawing on examples in research policy (Guston 2000a) and in environmental policy, including HEI (Keating 2001), agricultural extension (Cash 2001), and global climate change (Agrawala et al. 2001; Miller 2001), I argue that the presence of boundary organizations improves the context for the production of relevant knowledge and its application by decisionmakers and, moreover, that it does so while minimizing attendant risks of politicizing the science or scientizing the politics. The boundary organization offers an almost Madisonian solution in the reciprocal sharing and balancing of interests and accountability between politics and science.

This perspective is sympathetic to the vision of the position of policy analysis in Morgan and Peha's first chapter in this volume (2003). Morgan and Peha characterize policy analysis as a joint product of theories, facts, and other expert knowledge, with policy problems defined by decisionmakers. Most critical in their characterization is the interposition of policy analysis between the technical experts and decisionmakers. This interposition was also critical to the vision of science in democratic politics that Don K. Price (1965) articulated in his "spectrum from truth to power" in which the "estates" of professional and administrative practice applied knowledge, according to private and public rules, respectively.

The value of such institutions stands in contrast to a mechanism for the exchange of advice and analysis that some critics of OTA have called for, namely, more direct contacts between researchers and members of Congress.[5] This model of direct contact fails on at least three accounts to distinguish between advice or analysis, on one hand, and advocacy on the other:

1. The exchange in direct contact is likely to be private rather than public, and it would therefore suffer from apparent if not actual politicization. Such is the case, for example, with the conflict over the formulation of energy policy in the Bush–Cheney administration.
2. The exchange would not be subject to critical appraisal by peers and other concerned parties; it would therefore likely suffer substantively even in the event that it was impartially rendered. This logic is essential not only to the important role of peer review in scientific publications but also to the strong bipartisan support for using forms of peer review in regulatory science and in analysis for Congress, courtrooms (Berger 2000; Breyer 2000), and states (CSG 1999), as well as federal regulatory agencies (e.g., S. 746 in the 106th Congress).[6]
3. Similarly, individual researchers are likely to have some insight over narrow and near-term extensions of their work, but not over the broad array of potentially long-term societal consequences that would ultimately interest decisionmakers. Thus, a member of Congress is not likely to be able to rely on a single expert, or even a small sample culled by the member's staff, to provide analysis of the nontrivial implications of scientific and technical complexities.

Although it may be politically astute to take greater advantage of individual experts skilled as communicators than OTA did, the process behind the analysis needs to be conducted in a public and participatory way to achieve the appearance and actuality of both neutrality and rigor.

Conclusion

The history of OTA portrays an organization poised on the awkward boundary between politics and science, charged to provide technically oriented, unbiased foresight to a traditionally short-sighted, partisan, and particularized legislative body. Not surprisingly, the early history of OTA was shaky. Also unsurprisingly, this charge led to OTA's frequent and reflexive study of the practice of assessment and analysis, ultimately finding that nearer term policy analysis began to dominate its activity.

Discussions about a new mechanism for the provision of scientific and technical advice to Congress need to address a number of questions about the relation between the nature of that advice and the structure and performance of that mechanism. As this chapter and the preceding chapter have attempted to show, OTA's experience, its reflexive study of practice, and additional scholarship suggest a number of answers to those questions.

1. Who controls the agenda for inquiry? Congress does! When OTA attempted its own society-based priority-setting exercise under director Russell Peterson, Congress objected strenuously. Having Congress explicitly in control of the agenda-setting mechanism is critical.
2. How can partisan tensions be managed? OTA's Technology Assessment Board (TAB) generally managed partisan tensions well, and the active participation of a wide diversity of stakeholders in advisory committees can assist as well.
3. How can jurisdictional tensions be managed? A TAB-like body should manage the agenda and the choice of inquiries, but jurisdictional tensions may also be avoided through more particularized service that attends to some needs of individual members of Congress and not just to committees, their chairs, and their ranking members.
4. Which advisory, analytical, and assessment tasks will be undertaken? The choice of tasks should be discussed in parallel to the question of centralized or distributed structure. It may be that if the structure is distributed—inside or outside Congress—then more than one structure will be needed.
5. What is objectivity (or disinterestedness or neutrality) and how can it be achieved? For OTA, objectivity was informed neutrality, achieved by structured responsiveness to bipartisan and multijurisdictional clients and an expansive vision of stakeholder participation. Objectivity cannot be achieved by separation except at the expense of agenda control, salience, and, potentially, legitimacy.

6. How can timeliness and reader-friendliness be ensured? These were real problems for OTA. Timeliness may be at odds with structured neutrality and participation. Achieving reader-friendliness is not too hard for centralized institutions, but could be very challenging for a distributed one, particularly an external one. New technology could assist timeliness by speeding review and permitting more interactive telecommunications.
7. Will there be other products beyond reports? OTA briefed staff, fielded inquiries, and provided testimony and other informational services, but these tasks were secondary to the writing of reports. However, the report process, and not the document itself, was the most influential aspect of OTA's work. It is hard to imagine a smaller scale process making that claim, and in this context it is important to realize that the report is not the silver bullet of policy analysis. Nevertheless, smaller scale informational products more widely distributed to Congress would likely be popular and helpful.
8. Will reports emphasize context or options? OTA never quite resolved this question, but it seems reasonable to determine it on a project-by-project basis.
9. Will reports present options or make recommendations? OTA stressed options, but the folk wisdom about OTA is probably less accurate in its eschewing of recommendations than reality. More "distant" organizations (e.g., the National Academy of Sciences) might be more ready to make recommendations, but will they be taken seriously or will they damage the reputation of the organization? A new advisory organization could solicit comments to be published in reports from stakeholders and others, something like the comments that the General Accounting Office publishes in its reports.
10. How can the scope of reports remain within the intent and purview of the Technology Assessment Act? OTA continually had to refocus on technology, but because its agenda was set by Congress, it did not matter until its demise. Congressional agenda setting ensured a natural drift from foresight to nearer term policy analysis. Anything but a renewed OTA or agreement for more intensive use of the National Academy of Sciences might need authorizing legislation.
11. Can and will the participatory and analytical modes of assessment be reconciled? Yes, they can. Participatory technology assessment need not be the modus operandi of any new institution, although it might be helpful for it to be important among several methods, particularly for framing and scoping studies. Congress might enjoy hearing more directly from "real people."
12. Will new modes like constructive and real-time technology assessment be explored? Like participatory technology assessment, constructive technology assessment and real-time technology assessment should be explored as part of the suite of methods on which a new or renewed office would draw, depending on congressional interest and topics of interest. Such

techniques might be encouraged as more integrated ways of learning from and evaluating the R&D portfolio.

13. How will learning from the analysis or assessment best be promoted? OTA relied on its stakeholder participation and natural diffusion to create learning, but a new or renewed organization should be more aggressive about promoting learning, including using participatory methods, individual experts (members of advisory panels), and new media.
14. How will a new mechanism situate itself between experts and decision-makers? OTA relied heavily on staff, with guidance from experts and stakeholders on advisory panels and significant peer and stakeholder vetting of drafts. Much of the voluntary participation occurred because of congressional affiliation and is a natural outgrowth of congressional agenda setting.
15. Will there be direct interaction between members of Congress and experts, or mostly between staff and policy analysts? OTA relied significantly on its staff and their interaction with congressional staff. The direct involvement by members of Congress was generally limited to the (fairly constant) members of TAB. A new organization, however structured, could at least take greater rhetorical advantage of direct interaction between members and experts, but keeping experts "on message" would be critical.

In its brief survey the recent literature on assessment and analysis, this chapter has alluded to some of the answers that practitioners and scholars have found in the study of OTA and of other producers of science and technology policy analysis. More pertinently, the chapter has proposed that how the design of such an advisory mechanism answers these questions will go a long way in determining the quality of its analysis and the success of its organization.

Notes

[1]This principle is inspired by such perceptive observers of the role of knowledge in the democratic tradition as Dahl (1989), Ezrahi (1990), and Lindblom (1990).

[2]Clark began on this theme in a paper (Clark and Majone 1985), closely related to OTA's work, on the critical appraisal of science and technology policy analyses.

[3]I develop and apply this point in Guston (1997 and 1999). An impressionistic evaluation on these criteria might rate OTA moderate on actual impact, high on impact on general thinking and on learning by participants, and low on impact on learning by nonparticipants.

[4]Gibbons (1993) agrees that OTA's involvement of "the principal stakeholders and interested public in its work by use of advisory panels and reviewers, while retaining full responsibility for the finished product, has contributed to its level of credence and political acceptance and also its high standing in the technical community."

[5]For example, Newt Gingrich, the former Republican Speaker of the House who presided over OTA's demise, recently reiterated his opposition to OTA by suggesting

that direct contact between scientists and members of Congress would be preferable to staff-to-staff contact. Gingrich made these remarks at the Symposium on Allocation of Federal Resources for Science and Technology, hosted by the National Science Board for the release of a new draft report (NSB 2001), which among other recommendations, advocated the creation of "an appropriate mechanism to provide [Congress] with independent expert S&T review, evaluation, and advice."

[6]See also Chubin and Hackett (1990), Guston (2000b), Jasanoff (1990), and Smith (1992).

References

Agrawala, Shardul, Kenneth Broad, and David H. Guston. 2001. Integrating Climate Forecasts and Societal Decisionmaking: Challenges to an Emergent Boundary Organization. *Science, Technology, & Human Values* 26: 454–477.

Andersen, I.E., and B. Jaeger. 1999. Scenario Workshops and Consensus Conferences: Towards More Democratic Decision-making. *Science and Public Policy* 26: 331–340.

Andrews, Clinton J. 2002. *Humble Analysis: The Practice of Joint Fact-Finding.* Westport, CT: Praeger.

Bereano, Philip L. 1997. Reflections of a Participant–Observer: The Technocratic–Democratic Contradiction in the Practice of Technology Assessment. *Technological Forecasting & Social Change* 54: 163–176.

Berger, Margaret A. 2000. Expert Testimony: The Supreme Court's Rules. *Issues in Science and Technology* 16: 57–63.

Bimber, Bruce. 1996. *The Politics of Expertise in Congress: The Rise and Fall of the Office of Technology Assessment.* Albany: State University of New York Press.

Breyer, Stephen G. 2000. Science in the Courtroom. *Issues in Science and Technology* 16: 52–56.

Cash, David W. 2001. 'In Order To Aid in Diffusing Useful and Practical Information': Agricultural Extension and Boundary Organizations. *Science, Technology, & Human Values* 26: 431–453.

Cash, David W., William C. Clark, Frank Alcock, Nancy M. Dickson, Noelle Eckley, David H. Guston, Jill Jaeger, and Ronald B. Mitchell. 2003. Knowledge Systems for Sustainable Development. *Proceedings of the National Academy of Sciences* 100: 8086–8091.

Chubin, Daryl, and Edward Hackett. 1990. *Peerless Science: Peer Review and U.S. Science Policy.* Albany: State University of New York Press.

Clark, William C., and Nancy Dickson. 1999. The Global Environmental Assessment Project: Learning from Efforts To Link Science and Policy in an Interdependent World. *Acclimations* 8: 6–7.

Clark, William C., and Giandomenico Majone. 1985. The Critical Use of Scientific Inquiries with Policy Implications. *Science, Technology, & Human Values* 10(3): 6–19.

Council of State Governments (CSG). 1999. *A State Official's Guide to Sound Science.* Lexington, KY: Council of State Governments.

Dahl, Robert. 1989. *Democracy and Its Critics.* New Haven, CT: Yale University Press.

Dürrenberger, G., H. Kastenholz, and J. Behringer. 1999. Integrated Assessment Focus Groups: Bridging the Gap between Science and Policy. *Science and Public Policy* 26: 341–349.

Ezrahi, Yaron. 1990. *The Descent of Icarus: Science and the Transformation of Contemporary Democracy.* Cambridge, MA: Harvard University Press.

Farrell, Alex, and Jill Jaeger (eds.). 2003. *The Design of Environmental Assessment: Global and Regional Cases.* Washington, DC: Resources for the Future.

Gibbons, John H. 1993. Science, Technology, and Law in the Third Century of the Constitution. In *Science and Technology Advice to the President, Congress, and Judiciary*, edited by William T. Golden. Washington, DC: American Association for the Advancement of Science Press, 415–419.

Gieryn, Thomas F. 1999. *Cultural Boundaries of Science: Credibility on the Line.* Chicago: University of Chicago Press.

Global Environmental Assessment Project (GEA). 2000. *Annual Progress Report to the National Science Foundation for Academic Year 1999–2000*, available at environment.harvard.edu:80/gea/pubs/00prpt.html (accessed May 4, 2003).

Grin, J., H. van de Graaf, and R. Hoppe. 1997. *Technology Assessment through Interaction: A Guide.* Working Document 57, The Hague: Rathenau Institute.

Guston, David H. 1997. Critical Appraisal in Science and Technology Policy Analysis: The Example of Science, The Endless Frontier. *Policy Sciences* 30: 233–255.

———. 1999. Evaluating the First U.S. Consensus Conference: The Impact of the Citizens' Panel on Telecommunications and the Future of Democracy. *Science, Technology, & Human Values* 24: 451–482.

———. 2000a. *Between Politics and Science: Assuring the Integrity and Productivity of Research.* New York: Cambridge University Press.

———. 2000b. Regulatory Peer Review. Paper presented at the annual meeting of the American Political Science Association. September 2, Washington, DC.

———. 2001. Boundary Organizations in Environmental Policy and Science: An Introduction. *Science, Technology, & Human Values* 26: 399–408.

Guston, David H., and Bruce Bimber. 1998. Technology Assessment for the New Century. Working paper No. 7, Edward J. Bloustein School of Planning and Public Policy, Rutgers, the State University of New Jersey, New Brunswick, NJ, available at policy.rutgers.edu/papers/7.pdf (accessed May 4, 2003).

Guston, David H., and Daniel Sarewitz. 2001. Real-Time Technology Assessment. *Technology in Society* 23: 93–109.

Hill, Christopher. 1997. The Congressional Office of Technology Assessment: A Retrospective and Prospects for the Post-OTA World. *Technological Forecasting & Social Change* 54: 191–198.

Hörning, G. 1999. Citizens' Panels as a Form of Deliberative Technology Assessment. *Science and Public Policy* 26: 351–359.

Jasanoff, Sheila. 1990. *The Fifth Branch: Science Advisers as Policymakers.* Cambridge, MA: Harvard University Press.

Joss, Simon, and John Durant. 1995. *Public Participation in Science: The Role of Consensus Conferences in Europe.* London: Science Museum.

Keating, Terry J. 2001. Lessons from the Recent History of the Health Effects Institute. *Science, Technology, & Human Values* 26: 409–430.

La Porte, Todd M. 1997. New Opportunities for Technology Assessment in the Post-OTA World. *Technological Forecasting & Social Change* 54: 199–214.

Lindblom, Charles E. 1990. *Inquiry and Change: The Troubled Attempt To Understand and Shape Society.* New Haven, CT: Yale University Press.

Miller, Clark. 2001. Hybrid Management: Boundary Organizations, Science Policy, and Environmental Governance in the Climate Regime. *Science, Technology, & Human Values* 26: 478–500.

Morgan, M. Granger, and Jon M. Peha. 2003. Analysis, Governance, and the Need for Better Institutional Arrangements. In *Science and Technology Advice for Congress,* edited by M. Granger Morgan and Jon M. Peha. Washington, DC: Resources for the Future, Chapter 1.

National Research Council. 1993. *The Structure and Performance of the Health Effects Institute*. Washington, DC: National Academy Press.

National Science Board (NSB). 2001. *The Scientific Allocation of Scientific Resources.* (draft for comment, March 29), NSB 01–39. Arlington, VA: National Science Board.

Price, Don K. 1965. *The Scientific Estate.* Cambridge, MA: Harvard University Press.

Schot, Johan, and Arie Rip. 1997. The Past and Future of Constructive Technology Assessment. *Technological Forecasting & Social Change* 54: 251–268.

Sclove, Richard E. 1999. The Democratic Politics of Technology: The Missing Half. The Loka Institute, available at www.loka.org/idt/intro.htm (accessed May 4, 2003).

Smith, Bruce L.R. 1992. *The Advisers: Scientists in the Policy Process.* Washington, DC: Brookings Institution Press.

Smith, Bruce L.R., and Jeffrey K. Stine. 2003. Technical Advice for Congress: Past Trends and Present Obstacles. In *Science and Technology Advice for Congress,* edited by M. Granger Morgan and Jon M. Peha. Washington, DC: Resources for the Future, Chapter 2.

Social Learning Group. 2001. *Learning To Manage Global Environmental Risks*, Vols. 1 and 2. Cambridge, MA: MIT Press.

Vig, Norman J., and Herbert Paschen. 2000. *Parliaments and Technology: The Development of Technology Assessment in Europe*. Albany: State University of New York Press.

5

The European Experience

Norman J. Vig

Even before the Technology Assessment Act was passed by Congress in 1972, European scientists and officials demonstrated keen interest in the concept of technology assessment (Hetman 1973). Although the first proposals for an office similar to the U.S. Office of Technology Assessment (OTA) were introduced in the German Bundestag as early as 1973, it was another decade (1983) before the first European parliamentary technology assessment (PTA) agency was established in France. Since then, 15 other PTA units have been created, and for the past decade these bodies have been loosely associated as members of the European Parliamentary Technology Assessment (EPTA) organization.[1] A recent in-depth analysis of the first six of these PTA units (in Denmark, France, Germany, the Netherlands, the United Kingdom, and the European Parliament) concludes that they have successfully institutionalized technology assessment as a parliamentary function despite many institutional barriers and have developed a broad range of technology assessment methodologies that are proving useful to decisionmakers and the public (Vig and Paschen 2000). Indeed, these "little OTAs" may now provide useful models for the U.S. Congress if it decides to reestablish a smaller version of the former Office of Technology Assessment.

It is ironic that all of the initial European PTA units were inspired by OTA. Those who founded their respective agencies frequently visited Washington to observe OTA activities in the 1980s. Although it was evident that OTA was structured to fit the particular institutional environment of the U.S. Congress and could not be replicated in parliamentary systems in which parliaments occupied a much weaker and less independent constitutional position (Vig and Paschen 2000, 14–22), OTA nevertheless continued to inspire European

PTA development, and its reports were frequently used as reference points for PTA studies. The leaders of the PTA agencies were dumfounded when Congress voted to abolish OTA in 1995. It is incomprehensible to European parliamentarians that the leading democratic legislature in the world should no longer wish to have the kind of scientific and technological analysis that OTA provided.

The European PTA experience is perhaps most relevant in the context of this book in demonstrating that a wide range of institutional arrangements are feasible for conducting technology assessment and similar advisory functions. As will be seen in this chapter, each of the six original PTA units was structured quite differently to fit its particular constitutional and legislative environment. In one case (Denmark), constitutional limits did not allow establishment of additional bodies within Parliament; consequently PTA functions were housed in a new institution outside Parliament (the Danish Board of Technology). In the Netherlands, a similar desire to ensure the independence of PTA resulted in housing it in an institute of the Royal Academy of Arts and Sciences. In another case (France), the constitution limits the number of committees that can be established in either house of Parliament; hence a unique joint committee (*délégation*) of the National Assembly and Senate was devised as the formal agency to conduct technology assessments. The German Parliament wrestled with alternate institutional proposals for 16 years before finally agreeing on a model that complied with its strict parliamentary rules in 1989; in this case the PTA function is contracted out to an independent research institute but is closely supervised by a regular committee of the Bundestag.

The models adopted by the British, German, and European parliaments may provide the most useful models for the United States. The British Parliamentary Office of Science and Technology is now a permanent unit of the U.K. Parliament, and it carries out a variety of advisory functions under the direction of a nonpartisan oversight panel made up of members of both houses of Parliament and outside experts. The German Technology Assessment Board is in many ways modeled after OTA, but it is operated under contract by a major outside research institute. The Scientific and Technological Options Assessment Office of the European Parliament in Strasbourg is a small unit that primarily services requests from the 20 European Parliament committees and contracts all research out to universities and other research facilities throughout the countries of the European Union.

In addition to implementing different institutional structures, the European PTA agencies have experimented with a broad range of new methodologies for technology assessment, including several that involve substantial public participation (Vig and Paschen 2000, Chap. 11). Indeed, it is not unfair to say that they have reconceptualized the basic nature and purpose of technology assessment in ways that have enriched the original OTA model (Vig and Paschen 2000, 8–18, passim).

Each national European PTA unit defines its functions and goals somewhat differently from the others, and these variations appear to reflect deep cultural differences over the nature of political discourse, the societal implications of technology, and the role of the state and the public in shaping technological development (Vig and Paschen 2000, 22–28, 369–370, 380). These cultural differences also influence the selection of topics for research, the processes and methodologies used, and the ultimate results of technology assessments covering similar issues (Vig and Paschen 2000, Chap. 9–10). Some European practices may be more suited to American political, cultural, and epistemological assumptions than others. Any attempt to recreate a science and technology unit in the U.S. Congress must proceed within the limits of widely shared values and assumptions among members of both parties about the nature of science and technology and their role in U.S. society.

Institutional Models

Here is a brief summary of the institutional structures of European PTA units based on the six-case comparative analysis.[2] All of these bodies have small staffs averaging fewer than 10 people and budgets of $2 million or less as of 1997.[3]

France

The French Office Parlementaire d'Evaluation des Choix Scientifiques et Technologiques (OPECST) was established by law on July 8, 1983. It is unique among PTA organizations in that OPECST is actually a joint committee (*délégation*) of 16 members of Parliament (8 from the National Assembly and 8 from the Senate) who are responsible for conducting technology assessments on topics that are proposed by other parliamentary committees, political party groups, or any group of 60 representatives or 40 senators. After proposals are screened by the full committee, a *rapporteur* is appointed from among its members to gather information from public and private sources, sort out questions and issues, conduct further investigations if necessary, and prepare a draft assessment report (with the help of a small committee staff). Public hearings may be held to gather expert testimony, but the rapporteur frames the findings and conclusions and writes the report. The report is then submitted to the full committee for revisions or approval and published as an official parliamentary document. Reports often contain policy recommendations that are taken up by other parliamentary committees, but they are also widely circulated to the public.

Denmark

The Danish Board of Technology (DBT) was established by law in 1985 and made a permanent institution by legislation passed in 1995. Because the Danish constitution does not allow new parliamentary institutions, it was set up as an independent organization that is formally under the Ministry of

Research. The executive board is appointed by the minister of research after consulting the Committee on Research of the Parliament. It is made up of 11 members, with 4 nominated by the Ministry and 7 by various interest groups and government organizations. It has total control of its own budget and approves the annual technology assessment work plan, which is implemented by a professional staff (secretariat), headed by a director. From the beginning, the DBT has had a dual mandate to carry out comprehensive technology assessments and to further public debate and citizen participation on technological questions affecting society (in the Danish tradition of "people's enlightenment"). In practice the secretariat conducts both "expert assessments" through ad hoc multidisciplinary project groups hired by the agency and "participatory assessments," which involve members of the public. The DBT has developed and refined methodologies for public participation such as "consensus conferences" and "scenario workshops," which are now being used by other countries in Europe and throughout the world.[4] Although the DBT is not organizationally attached to the Danish Parliament (*Folketing*), it has a formal link to the Committee on Research. The DBT is required by law to meet with parliamentary committees and to respond to parliamentary requests for advice. Its studies and reports are regarded as useful by members of Parliament both in explaining scientific and technical issues and policy options and in informing them about public attitudes toward technologies.

The Netherlands

The Netherlands Office for Technology Assessment, as it was originally called, was created by executive decree of the Minister of Science and Education on June 17, 1986. In 1994 its name was changed to the Rathenau Institute (RI).[5] The institute, like the Danish Board of Technology, has no direct connection to the Dutch Parliament but is instead one of several institutes of the Royal Netherlands Academy of Arts and Sciences. This arrangement was adopted originally to ensure its independence from the government and its ministries, including the Ministry of Science and Education, which provides most of its funds. Because the RI advises government ministries as well as the Dutch Parliament, it is not strictly a parliamentary institution. The RI's broad research program is prepared every two years by a board of directors and submitted to the Ministry of Education, Culture, and Science and ultimately to Parliament for approval. However, it is the RI's job to define and implement specific projects within the broad themes of the program (e.g., access to information technology, ethical implications of biotechnology). The institute is primarily a planning and management staff for reports contracted out to other organizations, although some smaller studies are conducted in-house. Although Parliament has always been a major "client," much of the RI's work in recent years has revolved around organizing public debates on ethical and political questions related to science and technology. Like OPECST and the Danish Board of Technology, therefore, its function is not to provide expert advice so much as to frame and promote public and political awareness and discussion about major technological developments affecting society.

European Parliament

Scientific and Technological Options Assessment (STOA) was established as a unit of the European Parliament's Directorate-General for Research in 1987. In a manner somewhat reminiscent of OTA, STOA is overseen by a panel made up of one member from each of the 20 permanent committees of the European Parliament. The panel adopts an annual work plan of projects proposed by the committees. STOA's concept of technology assessment is similar to that of the former OTA, which directly inspired its creation (Vig and Paschen 2000). However, STOA contracts all of its projects out to universities and research institutions throughout the countries of the European Union and has no in-house research capabilities. Its function is thus one of project definition and management. Although it was heavily criticized in its early years for the uneven quality of its reports, STOA has since developed a clearer mission statement and stronger project management guidelines. It thus serves as a model for a "bare bones," contract-out technology assessment service.

United Kingdom

Like STOA, the U.K. Parliamentary Office of Science and Technology (POST) was heavily influenced by the OTA model (its first director, Michael Norton, was the science attaché at the U.K. embassy in Washington from 1982 to 1986). Its origins lie in a bipartisan delegation of members from both houses of Parliament to Prime Minister Thatcher in 1986. Thatcher, herself a chemist, welcomed the idea but declined to support parliamentary funding, suggesting instead that funds be raised from grant-giving foundations and companies, academic institutions, and individuals (including interested members of Parliament[6]). POST was thus initially set up in 1989 as an extra-parliamentary body, supported by a specially created foundation. In 1992 it was incorporated into Parliament on a trial basis. However, since April 2001 it has been established as a permanent official institution of the British Parliament. POST has always had a small staff (five to seven, plus a few visiting fellows or volunteer interns) but has been ingenious in developing a niche in the parliamentary landscape. It first made itself useful by producing short (four-page) briefing notes on current policy issues involving science and technology for parliamentarians but has expanded its role to include production of larger technology assessments and a wide range of activities with select committees in both houses of Parliament.[7] POST draws on the volunteer contributions of a wide network of outside experts but has limited funds for contracting out project work. Virtually all the writing of reports is done in-house by staff with professional expertise in specific areas.[8] All publications are reviewed by outside experts and stakeholders to ensure accuracy. Its reports are scrupulously objective and balanced to avoid any hint of partisan bias in the sharply divided atmosphere of Westminster. Organizationally, POST bears some resemblance to OTA in that it is supervised by a multipartisan board made up of 14 members from both houses, 4 non-parliamentary members from the science and engineering community, and the director. The board decides on work areas, responding to suggestions from individual parliamentarians, POST staff, and external organizations and to requests from parliamentary select committees.

Germany

The Office of Technology Assessment at the German Parliament (Büro für Technikfolgen-Abschätzung beim Deutschen Bundestag, or TAB) was established by the Bundestag in November 1989 after 16 years of debate and consideration by two Enquete Commissions. To avoid creating what some perceived to be a new bureaucracy within Parliament and to comply with the strict procedural rules of the Bundestag, a compromise was reached whereby (1) TAB is officially a non-parliamentary bureau supplied by an outside contractor (currently the Institute for Technology Assessment and Systems Analysis at the Karlsruhe Research Centre), which also conducts much of the research for the bureau; (2) TAB works only for the German Parliament; and (3) TAB reports to the Research Committee[9] of the Bundestag, which acts as its steering body. Topics for study can be proposed by other committees, but the Research Committee decides by resolution which projects are to be undertaken, and a rapporteur group within the committee works closely with TAB in formulating and monitoring projects. TAB and the Research Committee place high priority on the scientific and technical accuracy of their analyses. Some of the scientific work is subcontracted to outside experts, and drafts are peer-reviewed by advisory panels, but final reports are written by the TAB staff and director. The Research Committee must accept the reports before they are published as Bundestag printed papers. The German technology assessment reports are then discussed by other parliamentary committees and may become the subject of a proposed resolution and full plenary session debate. This provision for a formal response to TAB reports distinguishes the German system from those of other parliaments (Vig and Paschen 2000, 111).

Conclusions and Recommendations

Five conclusions may be drawn from the experience of the six parliamentary technology assessment agencies briefly described in this chapter:

1. Technology assessment (or comprehensive analysis of scientific and technological issues and relevant policy options) can be successfully conducted on a much smaller scale than that of the former U.S. Office of Technology Assessment, which at its peak had some 200 professional staff and a budget of more than $20 million.

2. Technology assessment can be adapted to a wide range of legislative environments and can take many different institutional forms, ranging from independent institutions loosely linked to parliaments (Denmark and the Netherlands) to official in-house advice and assessment offices (the United Kingdom), assessment offices provided by outside contractors (Germany), administrative units for managing outside project contracts (European Parliament), and a joint parliamentary committee that conducts its own investigations (France).
3. Technology assessment itself is a flexible concept that can serve different political and parliamentary purposes and can use many different methodologies; there is no need to adhere to the original OTA model. Some European PTA processes (especially those of POST, TAB, and STOA) are, however, conceptually and practically closer to those of the former OTA than others.
4. Despite numerous institutional reviews and changes of government, European PTA bodies have been able to avoid political controversy and partisan attacks such as those that terminated the OTA.[10]
5. If the U.S. Congress wishes to consider reestablishing a technology assessment or science advisory capacity, it should engage in "reverse learning" by studying the "little OTAs" of Europe (possibly including some of those not discussed here). That is, it could consider both different institutional arrangements and a different mix of functions and products than those that characterized the original OTA.

Some of the European organizational models are more suited to the U.S. context than others. The Danish and Dutch institutions are probably too loosely connected to the legislative process to serve congressional needs. Although there is some precedent for a joint congressional committee that oversees research (notably the old Joint Committee on Atomic Energy), the OPECST model would probably not be compatible with existing committee structures or members' current role definitions. However, the British, German, and European Parliament structures deserve careful scrutiny. If Congress wishes to establish a small in-house office with a high reputation for objective, balanced scientific advice to members and committees, it should study the U.K. POST. If it would prefer a more "distributed" approach via an office that contracts out and manages projects conducted at universities and other qualified research centers, it should look to STOA at the European Parliament. If it is interested in the possibility of contracting out the advisory and analytical functions to a single outside contractor such as a think tank or other nongovernmental organization, it should examine TAB at the German Parliament.

My own preference would be for a combination of the British and other European models: a Congressional Science and Technology Office (CSTO) modeled after POST (also drawing on the other European models to a more limited extent) and scaled up 5–6 times,[11] with an initial staff of 30–40 and an annual budget of about $10 million, some portion of which would support outside contracts to approved research institutions. CSTO's functions would be (1) to provide accurate, objective, nonpartisan advice on specific questions to individual members and committees of both houses; (2) to conduct smaller (3–6 month) assessment projects in-house at the request of committees; (3) to organize workshops, discussion groups, and consensus conferences involving scientific and technical experts, members of Congress and their staffs, stakeholder groups, and representatives of the lay public, for the purpose of defining emerging problems and broad policy alternatives; and (4) to contract out larger technology assessment projects to qualified outside institutions (with a completion time limit of no more than one year). CSTO would need something like the bipartisan joint committee that governed OTA to provide political cover and credibility.

Such a body could provide badly needed information, analysis, and advice to Congress on many complex, long-term scientific and technological issues that currently lack definition or are plagued by ideological and partisan arguments over good science and bad science. Though scientific evidence is usually uncertain and evaluation can never be entirely value free, a body such as CSTO could help to clarify what is known and what is not and provide more coherent intellectual frameworks for considering legislative policy options. It could raise the level of understanding on many pressing issues before Congress and help restore a sense of rational civil discourse. And it could do this for a cost of about 10 cents per U.S. taxpayer per year!

Notes

[1]Denmark, Finland, Flanders, France, Germany, Greece, Italy, Netherlands, Norway, Switzerland, the United Kingdom, and the European Parliament are full members of EPTA. Austria, Belgium, the Czech Republic, and the Council of Europe have units that are not primarily oriented toward their respective parliaments but also carry out studies for parliaments; they are associate members of EPTA. Links to these bodies, as well as information on recent EPTA activities, can be found on the EPTA website, available at eptanetwork.org (accessed May 6, 2003).

[2]See Chapters 3–8 of Vig and Paschen (2000) for details. These chapters were all written by the (then) directors of the PTA units.

[3]For personnel and budgets in 1997, see Vig and Paschen (2000, 12).

[4]See EUROPTA Project (2000), Joss and Durant (1995), and Vig and Paschen (2000, Chap. 11).

[5]It was named after G.W. Rathenau, who had chaired a committee on microelectronics in 1980 that recommended a technology assessment body.

[6]Prime Minister Thatcher herself made the first contribution.

[7]POST does not advise "standing" (legislative) committees, which are controlled by partisan majorities. "Select" committees oversee policy implementation and administrative efficiency in government departments.

[8]Currently these areas are health and medicine; environment and energy; physical sciences; engineering; and information technology.

[9]This is the Committee for Education, Science, Research, Technology, and Technology Assessment.

[10]There was a threat, however, to terminate the Danish Board of Technology as a cost-saving measure after the 2002 election. Intervention by members of Parliament reversed the decision.

[11]This is roughly in proportion to population differences, though a larger staff could easily be justified.

References

EUROPTA Project. 2000. European Participatory Technology Assessment: Participatory Methods in Technology Assessment and Technology Decision-Making (draft report, available at www.Tekno.dk/europta [accessed May 6, 2003]).

Hetman, Francoise. 1973. *Society and Assessment of Technology.* Paris: Organisation for Economic Cooperation and Development.

Joss, Simon, and John Durant. 1995. *Public Participation in Science: The Role of Consensus Conferences in Europe.* London: Science Museum.

Vig, Norman J., and Herbert Paschen (eds.). 2000. *Parliaments and Technology: The Development of Technology Assessment in Europe.* Albany: State University of New York Press.

Part III

Possible Institutional Models

6

Thinking about Alternative Models

M. Granger Morgan and Jon M. Peha

The six brief chapters that follow in this section of the book describe and discuss a number of alternative institutional mechanisms that might be used to provide balanced, nonpartisan science and technology analytical advice to Congress.

Why consider alternative models? A few members of Congress advised participants in the June 14, 2001, workshop not to spend much time in "academic exercises designing optimal institutional arrangements," but rather to work on getting some practical solution in place. This was good advice. Yet at the same time, Kingdon (1984) notes that in political settings such as Congress, successful decisions typically result from the simultaneous alignment in a "policy window" of three strands of rather independent activity: a political moment that makes action propitious; a set of good policy ideas that have been developed and widely discussed and have gained support among policy professionals; and a group of "policy entrepreneurs" who are prepared to advance and promote those solutions to relevant political decisionmakers at the critical moment. Thus, the idea of trying to produce an optimal institutional design is unlikely to be useful or successful because no proposal is likely to be implemented in precisely the form it is made. On the other hand, the idea of spending some time developing a menu of policy options is worthwhile because if and when Congress decides to act on securing improved mechanisms to obtain analytical advice, members and staffers will have a richer palette of ideas from which to assemble their solutions.

Accordingly, readers should take care not to view the models described in the chapters that follow as a menu from which only one design should be chosen, precisely in the form described. Rather, you should view them as laying out a range of design possibilities. The best and politically most feasible arrangement is likely to be some combination or merger of several of the ideas developed here.

The historical record reviewed in the preceding chapters suggests several issues that any successful institutional model will need to consider. These include the following:

- How will the process ensure that requests for analysis, including the ways in which the questions are framed, are as balanced as possible?
- How will the process ensure that the analyses performed are comprehensive, balanced, and objective?
- Who will have authority to request an analysis?
- What form will analyses take? How comprehensive should they be, and how fast should they be done? What should the work products look like?
- How will external advice, perspectives, and reviews be obtained from experts and stakeholders and incorporated into the work products?
- Should all reports be public documents (as with the Congressional Budget Office and the old Office of Technology Assessment [OTA]), private to members of Congress but releasable if desired (as with the Congressional Research Service [CRS]), or confidential for members and staff only (as with advice from legislative counsel)?
- If requests exceed the analytical capacity of the system, what mechanisms will be used for determining which requests are declined, and how will the system be protected politically when it must decline requests?
- How expensive will the proposed system be, and from where will the resources come? How will the system be protected during times of budget cutting?
- Given likely resource constraints, how will the system obtain needed technical expertise?
- How attuned can the organization be to the particular (and peculiar) needs of Congress?

Any successful model must deal with the problem of achieving and maintaining wide bipartisan and bicameral support. This requirement suggests that, independent of the specific institutional form, a bipartisan, bicameral committee should oversee the process and play an active role in vetting requests for analysis and the contents of final work products.

If the process of analysis is centrally funded (as the OTA was), there must be some mechanism to allocate the scarce resource, which from the point of view of the requesting entities, is a free resource. In the case of the OTA, the bipartisan, bicameral Technology Assessment Board had to approve all study requests from committees and tried to operate as a screening group to limit demands on the agency to reasonable levels. In this it was only partly successful. Sometimes a study would be mandated in a piece of legislation (which typically did not also provide incremental funding to cover its costs). Sometimes a powerful member would push hard for a study that OTA management thought was not of great importance but that political considerations dictated should be undertaken. There is no obvious single solution to the resource allocation problem. Although a centrally budgeted resource has some advantages, a case can also be made for having at least some of the resources for commissioning analysis tied to user committee budgets, perhaps to be used on a matching basis with centrally budgeted funds. However, the allocation of scarce study resources is an issue that will have to be addressed in any successful model. We provide some of our own views on this issue in Chapter 10.

In addition to the routine approval of studies and work products, a bipartisan, bicameral committee that oversees any analysis system should exercise the even more important function of ensuring long-term balance. Such a committee must remain attuned over time to the broad pattern of requests and work products. To borrow a phrase from Aaron Wildavsky (1979), when the message is unwelcome, "speaking truth to power" can sometimes be hazardous to the speaker. An analysis process that is widely perceived as balanced and objective can probably survive a few analytical products that anger powerful constituents. However, if, over time, important constituents begin to believe that there is a systematic bias or pattern in the questions being addressed or the way that results are framed, then problems will certainly ensue.

This observation carries two implications. First, any analysis process must continuously work to build widespread support among members on a bipartisan, bicameral basis, so that when conflicts arise about specific reports or when party dominance and policy preferences shift in Congress, support for the analysis institution remains firm. A wise bipartisan, bicameral leadership should balance the potential benefits of taking on a highly controversial topic against the potential harm that such an analysis might do to the analytical organization. In many such cases, adequate analytical input may be available from other sources. On the other hand, sometimes it is precisely when many other studies are being performed and reaching conflicting conclusions that Congress most needs an impartial, unbiased source to sort among them, clarifying which differences are due to scientific or technical uncertainties and which result from different value judgements or political preferences. When the political risks to an in-house analysis group are judged to be too high, the bipartisan, bicameral leadership can always suggest that a study requester commission the study from a highly credible outside group such as the National Academies.

Even if it is necessary to occasionally forgo performing a study on a strongly politicized, highly controversial topic, this should have little effect on the overall value of an analysis organization. Congress often most needs analysis on the hundreds of important issues involving science, technology, and public policy in which relatively few outsiders are engaged and few if any are performing careful, balanced analysis.

If studies are primarily commissioned by and serve the needs of committees, the visibility of the system within the broader congressional membership will be limited. How can general support among members be maintained and be ensured when power structures shift over time? There are probably several ways to address this need, but any model that fails to give this issue at least some attention runs the risk of falling out of favor as political conditions shift.

Access to specialized expertise will be a problem in any analysis system. Given the enormous breadth of issues that come before Congress, there is no way that a moderate-sized analysis organization can have on its regular staff all the different types of expertise that it will need. Whereas a variety of mechanisms are possible to develop the needed coverage, any successful model must have some mechanism for addressing this issue.

Models that involve commissioning analysis services from groups outside of Congress face complicated distributional issues, especially if the practice involves recurring acquisition, and either the resources or prestige involved become substantial. An important role for all members of Congress is to look out for the interests of their constituents as federal resources are distributed. Thus, although it might be desirable to have the quality, analytical track record, and neutrality of an outside analysis group be the only criteria for their selection to support Congress with analysis, in reality there will always be countervailing pressures—both to spread the resources and prestige among a number of different states and to choose groups that might be more philosophically compatible with powerful members' political persuasions and policy preferences. Any model that uses outside groups on more than an occasional basis will need to be designed to deal with these pressures.

Finally, there is the actual work product itself. The models described in the chapters that follow use different organizations and people to perform the analyses. However, all of these models face the common problem of ensuring wide expert and stakeholder input to the analysis process and wide review of draft work products.

Many critics of the OTA argue that its studies took too long. It might be argued that because many of the issues that OTA addressed were ongoing problems that would remain on congressional agendas in one form or another for many years, a year or two was not too long. However, that is beside the point. If many members believe that a successful analytical unit must supply its products more rapidly than the OTA did, that should be an important design constraint. There is a lot of room for alternatives between a CRS work product that is produced in just a few hours or days and an analytical product that takes a couple of years to produce. Much useful review and analysis can be done in periods of a few months. A successful model needs to be able to produce at least some of its products on that sort of time scale and should probably aim to limit most of its studies to under a year. Some complex topics might best be addressed with a series of closely linked studies that collectively run for longer than that. The key point is that the duration of the study should be negotiated at the beginning, and careful attention must be given to the needs of the congressional client. After an advisory organization has conducted one or more extended studies in a general area, it is often possible to put together a brief, narrowly defined report or presentation on a specific issue in a matter of days. Thus, one may not have to decide between doing only long-term work and only short-term work; the missions may be complementary.

There is obviously a trade-off among study duration, comprehensive treatment, and extensive external input and review. Some minimum level of external input and review is critical to ensure technical and political balance. That should probably set the minimum study time. Beyond this minimum time, the analytical process should be able to offer its clients two or three alternative work products that provide different levels of detail, breadth, and depth, with different resource levels and performance times. Given such a menu of study options, senior management and the bipartisan, bicameral committee that oversees the process must be willing to resist inevitable pressure to want more in less time.

OTA studies typically resulted in three products: a detailed report that often ran to 100 pages or more, a much briefer summary report of a few dozen pages, and a one-page summary. An adequate treatment of a complex issue, which acknowledges all important perspectives and stakeholder views, necessarily takes some space. However, it can certainly be done more concisely than was typical of the old OTA. In addition to agreeing up front on how long an analysis will take, an up-front agreement on the size of the work products should also be possible. Additional working materials and back-up documentation could always be made available via the Internet, in much the way that journals like *Science* and *Nature* now do routinely.

All six of the chapters that follow assume that the work products of the analytical process will be public documents. This assumption is motivated by two considerations. If the analytical process is to be seen as balanced and nonpartisan, it must be transparent. Observers must be able to independently judge the products it produces. If its products are primarily to inform the decisionmaking of committees, those inputs should be part of the public record. Congress makes most of its decisions through open, participatory, democratic procedures. However, one disadvantage of this constraint is that individual members have more incentive to protect an organization that serves as an extension of their own staff, and this is more likely to be the case when products are provided to members confidentially.

These, then, are some of the issues that any successful design must be able to address. As we noted above, readers should not read the chapters that follow as a search for the optimal institutional design. Rather, you may view them as a menu of ideas from which an analytically effective, politically feasible analytical system, or set of systems, might be assembled to provide better advice for the U.S. Congress on complex issues involving science and technology.

References

Kingdon, John W. 1984. *Agendas, Alternative and Public Policies.* Boston: Little Brown and Company.

Wildavsky, Aaron B. 1979. *Speaking Truth to Power.* Boston: Little Brown and Company.

7

An Expanded Analytical Capability in the Congressional Research Service, the General Accounting Office, or the Congressional Budget Office

Christopher T. Hill

This chapter explores the pros and cons of scenarios under which a capability like that of the Office of Technology Assessment (OTA) would be established as a unit of one of the existing legislative support agencies, namely, the Congressional Research Service in the Library of Congress (CRS), the General Accounting Office (GAO), or the Congressional Budget Office (CBO).

The chapter begins with overviews of the roles that these agencies have generally played in providing science and technology advice to Congress, compared with the roles that OTA played in serving Congress and the general public. It turns next to some scenarios for how science and technology policy advice might be organized within each of the agencies. To provide a simple framework for analyzing the possibilities of alternative arrangements, attributes that are essential for a science and technology advisory body to serve Congress and the public are enumerated. Comparing the attributes of the three existing agencies with these essential attributes provides the basis for the conclusion that such an arrangement is unlikely to be successful.

The Roles of Science and Technology Advice in Congress

In addition to such questions as the nature of the work that a scientific advisory body in or for Congress should do, as well as where it might be located, it is important to consider carefully the roles that OTA played and that any new agency offered in its place might be expected to fulfill.

Other chapters in this volume seek to specify the content and nature of the work products of a scientific and technological advisory body for Congress. Such a body should be able to address a wide range of public policy issues in which scientific and technical understanding and/or technological applications are important to generating the issue, illuminating the consequences of private and public actions, or addressing problems of society that present themselves to Congress for resolution.

In considering possible new ways to organize science and technology advice for Congress, it is important to keep in mind that OTA was the outcome of a protracted debate about the need to create a "technology assessment" function somewhere in the U.S. governance system. The establishment of OTA as an advisory body to Congress was driven only in part by its role in providing science and technology advice to Congress itself. OTA was intended by its proponents to serve both the executive and legislative branches of government as well as the polity at large. The greatest value of OTA may have been its contributions to structuring the broad public debate on important issues, rather than its contributions to informing the congressional process per se. In a sense, Congress served OTA and, through it, the public, even as OTA served Congress. Congress provided three key things to OTA: a demanding client, effective access to key interest groups from every quarter, and the funds to do its work (Hill 1997).

Thus, the ability to help structure the larger public dialog on key public policy matters is an essential characteristic of any new scientific and technical advisory body designed to serve Congress. Fulfilling this role demanded of OTA an uncommon degree of openness and inclusion of the public, including organized interest groups, as it did its work. Whereas the advocates of "public participation" were often frustrated by what they perceived as inadequate attention to public input, OTA nevertheless was considerably more open to public input than any of the other congressional support bodies.

Science and Technology Advice in CRS, GAO, and CBO Today

Congressional Research Service

The Congressional Research Service, then the Legislative Reference Service, established the Science Policy Research Division (SPRD) as a separate unit devoted to science policy matters in 1965. This development reflected the growing importance of scientific information to national policymaking in such areas as the space program, national security, mathematics and science education, the technological effects of the Vietnam conflict, and the then-emerging environmental movement. Since that time, CRS has employed some 20–75 staff working in related policy areas. (The exact number of science and technology policy staff working at CRS depends on the time period and on how one categorizes the fields of expertise and activity of the CRS staff in various administrative units.) As a result of a recent reorganization of CRS, the SPRD ceased to exist as an independent unit. The SPRD staff concerned with health matters were assigned to what is now the Domestic Social Policy Division, and most of the others became part of the Resources, Science, and Industry Division of CRS.

The CRS staff represent a wide range of expertise in scientific and technical matters. Only a few of the science policy experts at CRS have had advanced education or experience in scientific or engineering fields; the academic expertise of many is in science policy as a field of policy studies or in diverse fields of the social and policy sciences and humanities. During the late 1980s and the 1990s, CRS has focused its staff hiring at the more junior levels of the civil service system and did not attract experienced scientists and engineers with advanced training to the staff. At one time, the rank of "senior specialist" at CRS was frequently used to hire such persons; in more recent years, the senior specialist ranks have been filled from within or by former division chiefs, rather than by senior hires from outside. Owing to a change in recruiting practices, since 2001 CRS has hired a number of doctoral-level, experienced scientists and engineers.

CRS was significantly involved in the creation of OTA. In the late 1960s and early 1970s, senior staff of the Legislative Reference Service (LRS, later CRS) played key roles in the discussions leading to the introduction of the Technology Assessment Act of 1972. As part of that process, LRS prepared the influential report, *Technical Information for Congress* (LRS 1969), based on a set of case studies that illustrated the problems of providing technical information in the legislative context. One of the circumstances leading to the establishment of OTA was the recognition that CRS, at least at that time, was unable to provide the kind of expertise and institutional structure required to fulfill an OTA-like mission. Evidence of the perceived relevance of CRS to OTA was the provision of the Technology Assessment Act of 1972 that made the director of CRS and the comptroller general of the Unites States ex officio members of the Technology Assessment Advisory Council of OTA. Staff of SPRD provided support to the CRS director in this role.

CRS is renowned for its ability to provide Congress with highly targeted information, including technical information, on very short notice and in forms highly tuned to the congressional decisionmaking process. On the other hand, CRS makes limited use of external sources of information and even less of external analysis. Its approach tends to be reportorial rather than analytical. It tends to present the views of all interested parties as if they were of comparable validity, rather than to analyze each such view skeptically so as to arrive at the best available understanding of an issue, as one would expect from a new analytical capability.

CRS produces a wide range of products for Congress, ranging from one-sentence answers to specific questions to occasional book-length studies. The bread and butter of the CRS line of work products includes issue briefs of 10–20 pages that summarize the state of congressional activity on specific issues, background papers for committees to use in preparing hearings, and collections of published literature intended to illustrate diverse contemporary perspectives on current issues. CRS answers literally hundreds of thousands of requests from members, staff, and committees of Congress each year. CRS also serves Congress as its main institutional memory regarding the activities and decisions of Congress itself.

The CRS culture is "internalist"; that is, CRS analysts for the most part draw their knowledge and analytical base from other groups in government, typically the executive branch agencies. Much of their expertise on any subject has to do with what government is doing about it, rather than with the scientific basis for understanding it.

CRS operates under strict rules protecting the confidentiality of its work for Congress and even, for some products, the identity of the specific congressional clients for its work. This closeness tends to freeze out external input. Furthermore, for many years Congress has expressly prohibited CRS from publishing most of its reports for the general public, which has sharply attenuated the responsiveness of the agency to inputs and assessments from those outside its walls. As a graphic illustration of the degree to which CRS is not open to direct communication with the general public, it does not maintain a public website, even within the Library of Congress's Thomas system. Its reports, its organization, and its staff list are not available to the general public on line.

General Accounting Office

The General Accounting Office (GAO) is one of the oldest support agencies for Congress, established pursuant to the Budget and Accounting Act of 1921 to improve federal financial management. The same act established for the first time a uniform and comprehensive budget for the federal government. GAO largely "looks back" to determine whether the public is being well served by public programs and expenditures. Here is how GAO portrays itself on the top page of its website:

> The General Accounting Office is the investigative arm of Congress. GAO exists to support the Congress in meeting its Constitutional responsibilities and to help improve the performance and accountability of the federal government for the American people. GAO examines the use of public funds, evaluates federal programs and activities, and provides analyses, options, recommendations, and other assistance to help the Congress make effective oversight, policy, and funding decisions. In this context, GAO works to continuously improve the economy, efficiency, and effectiveness of the federal government through financial audits, program reviews and evaluations, analyses, legal opinions, investigations, and other services. GAO's activities are designed to ensure the executive branch's accountability to the Congress under the Constitution and the government's accountability to the American people. GAO is dedicated to good government through its commitment to the core values of accountability, integrity, and reliability. (GAO 2003)

In fulfilling its mission, GAO from time to time issues reports on the activities of the federal agencies and programs that support science and technology, as well as on issues in which scientific and technical information are central. However, its focus is typically not on the implications of scientific discoveries or technological activities but instead on government program implementation and finances. The bulk of GAO's work is retrospective; that is, GAO examines the implementation and consequences of federal policies and programs that have already been put into place. Although it has diversified in recent decades, the fundamental ethos of the agency is that of the accounting and auditing professions—GAO looks back, carefully. GAO has not placed much emphasis on policy analysis and on anticipating the needs for and consequences of new federal policies.

The GAO analytical process is closely held, with inputs from agencies limited to published commentary on the draft GAO reports and to information requested by auditors. One of the strengths of GAO is its field offices located around the country whose staff can be called on to obtain data at remote locations.

Because GAO operates from the ethos of the accounting and auditing professions, it tends to discourage the development and use of specialized subject-matter expertise on its staff. Under this operating premise, staff development of deep expertise in an agency, issue, or technology would put the objectivity of the staff at risk. Hence, there is limited opportunity for GAO to develop a staff that is expert, technically based, and skilled in advanced analytical methods and approaches applicable to science and technology policy issues.

Under Comptroller General Elmer Staats, who served from 1966 to 1981, the General Accounting Office established a science policy staff unit. This unit reflected Staats's personal interest in science policy matters, nurtured by his experiences in the post–World War II Bureau of the Budget, where he helped to establish a variety of federal science agencies. It also supported the comptroller general in fulfilling his or her responsibility as a statutory member of the OTA Technology Assessment Advisory Council. After Staats's departure and the retirement of the leader of this group, Osmund Fundingsland, the form and composition of this dedicated science and technology policy staff evolved, and it was eliminated some time ago as a separate entity. Today significant technical expertise resides in a unit called the Applied Research and Methods Group.

Congressional Budget Office

The third of the surviving analytical support agencies for Congress is the Congressional Budget Office (CBO), established in 1974 under the Budget Impoundment and Control Act. As stated on its website, the mission of CBO is "... to provide the Congress with the objective, timely, nonpartisan analyses needed for economic and budget decisions and with the information and estimates required for the Congressional budget process" (CBO 2002).

Of all the support agencies, CBO is by far the most committed to analysis based on (and therefore biased by) a single discipline, the field of economics. This is appropriate to the main business of the organization—analysis of the budgetary consequences of proposed public policies—but it leaves the institution culturally unsuited to wide-open inquiry into the possible future consequences and management of technological applications or to the provision of scientific and technical advice to Congress.

Like the other support agencies, CBO is "internalist" in focus, tending to base its analyses on matters internal to the functioning and budgeting of federal programs and policies. Other than its deep commitment to analytical economics, both micro and macro, it draws in only a limited way on the expertise of people versed in science and technology.

In keeping with its limited role in program and policy analysis, science and technology policy matters have always been of marginal interest to the Congressional Budget Office. Under an early deputy director for programs, David Bodde, an engineer, CBO examined a range of budget and program issues in science and technology. For at least the past decade, however, CBO's program interest in these fields has been vested largely in a single economist, who is on the staff of the Microeconomic and Financial Studies Division of CBO. He has prepared a series of studies of major federal science and technology programs focused on, but not entirely limited to, their budgetary implications.

Essential Attributes of a Science and Technology Advisory Apparatus for Congress

Based on experience with the Office of Technology Assessment and the other congressional support agencies, there are a number of desirable attributes for any organization that wishes to provide authoritative, informed, unbiased, and timely information and advice to Congress on important public issues involving science and technology. Such an organization should be:

- technically expert—staff must be able to assess and judge scientific and technical information from diverse fields, to locate the nation's experts quickly, and to converse with such experts on the basis of a high level of mutual expert understanding;
- analytically competent—the staff must be highly competent in modern methods of policy analysis such as applied economics, legal analysis, and mathematical modeling and simulation;
- inquiring—the staff and the procedures used in the work of the organization must assume an analytical stance that accepts no externally supplied information from a single source as being true or complete, and they must constantly inquire into the foundations of any claim made about a technology or its consequences;
- politically wise—the staff must understand the nature of the congressional client, especially regarding the levels of analysis, data, and communication that can be effectively integrated into the congressional decisionmaking process;
- responsive—the organization must be responsive to congressional expectations for the issues to be addressed, the time frame for analysis, and the

scope of projects, and it must be equally responsive to the entreaties of parties at interest who wish to provide input into its studies;
- open and inclusive—the organization must be open to ideas and information from all credible and relevant sources;
- unbiased—the organization must seek to provide analysis that is not biased by virtue of the political perspective of its participants, the disciplinary backgrounds of the analysts, or the methods used in the analysis;
- dynamic—the organization and its staff must be able to address a wide variety of policy issues and underlying bodies of scientific and technical understanding, and it must have the flexibility to adjust its topics of study in rapid response to changing external circumstances;
- focused—the organization should be focused on the few important functions and tasks that it performs best and should not be distracted by opportunities or requests to perform related but ultimately less valuable functions because it has the resources to do so;
- credible—the organization must be seen and experienced within Congress, by expert communities, and by the general public as credible and believable; and
- circumspect—the organization and its staff must respect the needs of members, committees, and staff of Congress for it to act in such a way as to bring credit to Congress as a whole and to its members.

Fulfilling these 11 attributes in sufficient measure would be a Herculean task for any organization. In my judgement, OTA did well on most of these and superbly on several. Undoubtedly, other observers might well identify additional indispensable attributes of a science and technology advisory body for Congress.

Organizing Science and Technology Policy Advice in CRS, GAO, or CBO

This section offers scenarios for how a science and technology policy advisory unit might be organized within each of the existing congressional support agencies. These scenarios, which are intended to represent an optimistic view of each agency's potential to play this role, are evaluated in a later section.

Congressional Research Service

Within the Library of Congress, CRS is the most likely home for a special unit devoted to science and technology policy analysis, assessment, and advice. From a substantive perspective, such a unit could draw on the many subject-matter experts at CRS, as well as on its review and publications arms. The former Science Policy Research Division in CRS might have offered a suitable organizational unit within which to house a science and technology advisory unit for Congress on the model of an OTA, although that option is no longer available. Organizationally, the unit could be set up separately, equivalent to a research division, or it could be given special status as an office reporting directly to the CRS director. In theory it could function as a "virtual" division, drawing from all parts of CRS without its own organization or staff, but this mode of organization seems highly unlikely in light of the myriad competing demands on the time of the CRS staff.

Alternatively, the Library of Congress could house an OTA-like enterprise separate from CRS as a unit of the library equivalent to CRS in formal stature, similar to the Federal Research Division, which does contract research for executive branch agencies in national security areas.

General Accounting Office

GAO could establish a new science and technology advisory unit located in one of its "teams" (divisions) or attached to the comptroller general's office. To fulfill a science and technology advisory mission, GAO would have to hire new staff from outside and upgrade the policy–analytic skills of existing staff. As currently structured, none of the teams seems to have the mission or organization to address complex issues involving technology in a prospective manner.

In debates leading up to the passage of Public Law 107–68, providing appropriations for the legislative branch for fiscal year 2002, Senator Jeff Bingaman (Democrat, New Mexico) introduced an amendment into the Senate version of the bill that would have provided GAO with $1 million in FY2002 for a pilot program in technology assessment. In a colloquy on the floor of the Senate, Bingaman indicated that this pilot program would be given the task of contracting with outside agencies such as the National Academy of Sciences to "utilize technology assessment methodology to analyze current science and technology issues affecting our Congress" (Bingaman 2001). He also expressed a preference for CRS to be given this responsibility but indicated that he would accept GAO in that role and would hope to change the assignment back to CRS in conference. The version that emerged from conference contained one-half million dollars for GAO to perform a technology assessment of the use of biometrics for border security. This assessment has now been completed (GAO 2002). To obtain an independent evaluation of this effort, GAO funded a small external evaluation committee to track the activity and provide an independent review of the undertaking.[1] Their conclusions read as follows:

> Given the relatively short time that was available to perform the assessment, and the fact that GAO has not previously engaged in such analysis, they have done a very good job. They successfully identified, described, and evaluated a range of relevant technologies. With assistance from the NRC [National Research Council], they also engaged, and in their report have reflected, the views held by experts, affected parties, and concerned stakeholders so that congressional staff and members can accurately identify and weigh the trade-offs involved in policy choices.

This first GAO effort has been rather less successful in framing the analysis broadly in such a way as to address the full range of issues which the Congress is likely to have to consider and in providing analytically informed input that will support the needs of congressional staff and members as they refine and tune legislative products. These limitations result from the fact that, while GAO has staff with excellent technical credentials, they have relatively little previous experience in framing and performing policy analysis. In addition, current internal GAO guidelines are not entirely consistent with the needs of technology-focused policy analysis and assessment.

If Congress concludes that it needs expanded technology assessment and related analytical capability to fill the gap between the short-term services provided by the CRS and the large-scale long-term studies of the NRC, it would be well advised to consider a range of alternative institutional mechanisms, of which studies conducted through the GAO are just one possibility. The problem should not be viewed in terms of choosing just one mechanism. It might be best to pursue several different strategies in parallel. The key point is to find better ways for Congress to obtain the types of assessment and analysis it needs in a timely manner so that it can be better informed as it addresses complex problems in technology and public policy.

While it will pose a significant challenge, we believe that if the Congress chooses to ask the GAO to perform additional technology assessments in the future, the problems we have identified can be resolved if GAO acquires more policy–analytic staff capabilities, makes appropriate changes in internal guidance and administrative arrangements, and makes effective use of outside expertise and contractors (Fri et al. 2002).

Congressional Budget Office

As in the case of GAO, CBO could organize a science and technology advisory unit on its program analysis side, but even more than GAO it would have to hire an entirely new staff to fulfill such a mission. Only a handful of the current staff would appear to have backgrounds appropriate for a science and technology advisory unit.

Do the Three Congressional Support Agencies Have the Attributes Essential for Success?

In my judgement, none of the three agencies—CRS, GAO, or CBO—has the essential attributes identified above to be an ideal host for a new science and technology policy advisory mechanism. Of the three, CRS offers the greatest promise but it, too, has limited potential.

To address whether the agencies have the attributes for success, Table 7-1 illustrates my assessment of how each agency measures up on each of the

Table 7-1. An Assessment of Each Support Agency on the Attributes for Success

Attributes for success	*CRS*	*GAO*	*CBO*	*OTA*
Technically expert	medium	medium	very low	very high
Analytically competent	low	low	high	high
Inquiring	low	medium	medium	high
Politically wise	high	medium	very high	high
Responsive	very high	medium	high	low
Open and inclusive	low	low	very low	very high
Unbiased	high	medium	low	medium
Dynamic	high	high	low	low
Focused	low	medium	high	high
Credible	high	medium	high	high
Circumspect	high	medium	high	medium

attributes. Readers may well have different or supplementary views of the attributes of a successful organization. Also, of course, readers may have a different view of the "score" of each agency on each criterion. To provide a point of reference, I have included my assessment of the position of the former Office of Technology Assessment on the same criteria.

The qualitative assessments shown in Table 7-1 are based only on my own subjective experience and my sense of the views of others on these matters. They are offered here as a stimulus to discussion rather than as definitive statements.

Today, none of the existing congressional support agencies has the staff, the financial resources, the operating procedures, the interactions with the external world of expertise and experience, or the broadly based analytical competence to take on the OTA roles. Their cultures are in large measure, although differently, antithetical to a competent science and technology policy advisory body of the scope contemplated for this important project.

CRS is highly responsive, circumspect, and respectful of congressional prerogatives. However, it has only moderate technical and analytical resources to carry out a science and technology advisory operation in the OTA style. Furthermore, it has none of the tradition of openness to the public and to interest groups that is necessary to ensure that its products are unbiased and of the highest quality. Instead, it depends on internal staff quality control for these purposes.

GAO remains firmly rooted in the auditing tradition, eschewing both subject matter expertise and openness to external contributions and review. Its policy–analytical capabilities are severely limited, and its scientific and technical staff is modest. GAO has recently piloted a technology assessment that was moderately successful. However, if it were going to continue to serve this function, it would need to significantly upgrade its staff capabilities and address some fundamental issues of instructional culture. Furthermore, GAO is often used in its traditional role by a single member of Congress, a small group of members, or a committee in pursuit of specific political objectives related to finding fault with government programs or agencies. In such an environment, the kind of objective analysis implied by an OTA function would not thrive. As Fri et al. (2002) have noted,

> If the Congress is going to make regular use of the GAO as a vehicle for technology assessment, it should give some thought to how to provide bipartisan bicameral oversight. Such oversight is important to assure that the assessment function:
> — remains responsive to the needs of the Congress;
> — is not unduly influenced by the preferences and agendas of particular members, political groups, or parties;
> — maintains a strict neutrality on issues of public policy; and
> — uses scarce resources on problems that are of the greatest importance to the Congress and the nation.

CBO, appropriately, is focused on budgetary and financial management implications of congressional decisionmaking, especially around program funding. Broadly based issue analysis is not its strength. The world view of the discipline of analytical economics dominates the agency's leadership and staff, and it is difficult to see how CBO could overcome this style to host a scientific and technical advisory body.

Conclusion

Congress and the country desperately need a strong capability to analyze important public policy issues in which scientific and technical understanding play central roles. The former Office of Technology Assessment offered a highly successful model of how to organize and manage such an agency within the congressional context.

Today, none of the three surviving congressional support agencies—CRS, GAO, or CBO—has the desired mix of attributes in sufficient measure to fill the vacancy in the national policymaking system caused by the demise of OTA. GAO and CBO are structured around specific styles of professional practice in auditing and economics that do not fit the model of an OTA. The recent experiment conducted at GAO suggests that technology assessment might be developed there, but substantial staff and institutional changes would be required if it were to succeed. CRS is a much more eclectic agency than CBO or GAO, but it is focused on close support of individual members, staff, and committees of Congress, with none of the commitment to high-level public input that is necessary to ensure quality and to create a sense in the communities at interest that they have had the opportunity to be heard and respected. Nothing in the contemporary record of these agencies suggests that they could be flexible enough to accommodate an OTA in the near term.[2] Creating an effective analysis and assessment capability in any of these agencies would pose large, perhaps insurmountable, difficulties.

Notes

[1]The Fri et al. (2002) report is reproduced in Appendix 3.
[2]For a more optimistic view, see Appendix 3.

References

Bingaman, Jeff. 2001. *Congressional Record.* pp. S8008–S8009, July 20.
Congressional Budget Office (CBO). 2002. www.cbo.gov/about.shtml (accessed February 10, 2002).
Fri, Robert W., M. Granger Morgan, and William A. (Skip) Stiles Jr. 2002. *An External Evaluation of the GAO's Assessment of Technologies for Border Control.* Washington, DC: General Accounting Office.
General Accounting Office (GAO). 2002. *Technology Assessment: Using Biometrics for Border Security.* GAO-03-174. Washington, DC: General Accounting Office.
———. 2003. www.gao.gov (accessed May 11, 2003).
Hill, Christopher T. 1997. The Congressional Office of Technology Assessment: A Retrospective and Prospects for the Post-OTA World. *Technological Forecasting and Social Change* 54: 191–198.
Legislative Reference Service (LRS). 1969. *Technical Information for Congress.* Committee Print of the Committee on Science and Astronautics, U.S. House of Representatives, April 25, 1969.

8

Expanded Use of the National Academies

John Ahearne and Peter Blair

In one form or another the units of what now make up the National Academies have been supplying advice to Congress since 1863, when Congress chartered the National Academy of Sciences to "whenever called upon by any department of the Government, investigate, examine, experiment, and report upon any subject of science or art." This charter was signed by President Lincoln during the height of the U.S. Civil War, and the president was the first to call upon the academy for advice. This first request was to help design a compass that would work in ironclad warships.

Today, the National Academies refers to a collection of four units: the National Academy of Sciences (NAS), the National Academy of Engineering (NAE), the Institute of Medicine (IOM), and the National Research Council (NRC). The academies and the institute are honorary societies that elect new members to their ranks each year.[1] The National Research Council, with more than 1,100 staff, is the operating arm of the National Academies, which has carried out studies for Congress and federal agencies since its founding in 1916. In this chapter we briefly examine the evolution of the structure and operations of the NRC and explore the prospects for creating a mechanism by which Congress might make more regular and systematic use of the National Academies. The first part of Appendix 2 provides some additional detail regarding the history and mission of the National Academies.

NRC Advice to Congress: The Current Model

Consistent with the NAS's charter, the NRC principally carries out studies for the federal government, although predominantly for executive agencies. However, Congress also frequently mandates studies by the NRC, either specified in law or directed in report language accompanying legislation. In the period between the 102nd and 106th Congresses (1991–2000), there were congressional mandates for 186 studies. There were 40 mandates in the 102nd, 25 in the 103rd, 22 in the 104th, 53 in the 105th, and 46 in the 106th. To illustrate the scope of these mandates, the second part of Appendix 2 includes a listing of the titles of congressionally mandated studies carried out by the NRC between 1991 and 2000. The NRC also performed many more studies directly for federal agencies.[2] For our purposes, however, we refer only to those studies done as a result of direct legislative requests either signed into law or directed in report language accompanying the legislation.

Many of these mandated studies were health-related, such as examining the health consequences of allowing for higher levels of copper in drinking water. Some were for specific technological issues, such as the engineering design of the international space station, and others were quite broad, such as strengthening science at the U.S. Environmental Protection Agency or building a work force in the information economy. The congressional Office of Technology Assessment (OTA) was closed in the 104th Congress, and 53 mandates for the NRC and IOM subsequently emerged in the 105th Congress. It appears that the number of congressionally mandated NRC and IOM studies jumped from approximately 30 per Congress to about 50 per Congress after the demise of the OTA. James Jensen, who was OTA's congressional liaison and now directs NRC's Office of Congressional and Governmental Affairs, believes based on his experience in the transition that some but not all of the recent increases in NRC and IOM congressionally requested studies have resulted from OTA's absence. The National Academies have benefited from hiring many former OTA staffers.

Why does Congress turn to the NRC? The following sections describe many of the strengths commonly cited when Congress turns to the NRC as well as weaknesses of the NRC's structure, process, and traditions in carrying out studies for Congress. We write this with some hesitation, however, because in an attempt to be organized and simple, it is sometimes easy to drift into the simplistic. In fact, the NRC has a wide range of traditions and variations in its processes to accommodate the nature of the subjects it addresses and its various sponsors, although we will attempt to address the core process at the heart of all studies.

Traditionally Cited Strengths and How They Are Being Exploited

The key strengths of the NRC in providing advice to government, and especially to Congress, are its long-established reputation for credibility, its convening power, and the integrity of its study process (see National Academies 2000, 2002). Some features of these strengths include the following:

Credibility. Perhaps the principal strength of the NRC is its credibility, largely enabled by its association with the NAS, NAE, and IOM. The process by which this nongovernmental institution conducts its work is designed to ensure its independence from potential outside influences and political pressures from government officials, lobbying groups, and others. The phrase "a prestigious National Academy study showed," which most often refers to an NRC study, is typically the lead to a news story covering a new report. Many of the recommendations in NRC studies are directly and often immediately implemented by federal agencies. Sometimes they are incorporated into legislation.

Convening Power. A second major strength is the convening power of the NRC, that is, the people who participate in the studies. Most studies are carried out by groups of volunteers who are broadly considered among the best experts on the issues to be studied, including members of the National Academies as well as many others. Although the issues under study by NRC committees are most often centered on the United States, committee membership often includes experts from overseas as well. For example, a study released in 2001 (NRC 2001a), completed for the U.S. Department of Energy, examining geologic disposition of high-level radioactive waste, used a committee whose membership included scientists from Belgium, Japan, Sweden, Switzerland, and Russia.

Because of the breadth of membership in the NAS, NAE, and IOM and the links of the organization to the scientific and technical communities worldwide, the NRC is well equipped to identify the leading experts to serve on study committees. Although the committee members are not compensated for their work (travel expenses are paid), the track record of NRC efforts shows that it has not been difficult to engage distinguished and respected committee members. Service on an NRC committee is seen as an honor as well as an important public service.

Study Process. Finally, another key strength that has continued to evolve over the years is the study process itself, which is designed to maintain balance and objectivity. For example, after consensus is achieved by a study committee and a draft report is prepared by the committee or with staff support on behalf of the committee and endorsed by the committee as reflective of the group's deliberations, findings, and conclusions, the report is circulated for peer review. Typically 8 to 12 peer experts, anonymous to the committee, are asked to serve in this capacity. The NRC process requires the committee to address all of the comments from these reviewers. Furthermore, to help ensure that the committee considers the reviewers' comments in a comprehensive manner, the NRC appoints a monitor, who is responsible to the NRC's Report Review Committee, who must decide whether the study committee has sufficiently addressed all of the reviewer comments. The monitor can and often does return proposed responses to the study committee for further work. This process, while continuing to evolve, has proven effective over the years in maintaining the quality of NRC reports.

Commonly Cited Weaknesses and How They Are Being Addressed

The most common criticisms of the NRC study process are the following:

Cost. The NRC is generally considered a high-cost operation. Even though committee members are volunteers whose time is contributed pro bono (travel expenses are covered by NRC projects), the overhead for the NRC is necessarily substantial, partly because many of the staff supporting studies are professionals and partly because supporting the infrastructure necessary to maintain access to key sources of volunteers, including the governance structures of the National Academies, must be maintained. Perhaps two American aphorisms, "there is no free lunch" and "you get what you pay for," apply here because ensuring access to the best and brightest pool of experts and ensuring an objective and authoritative study process together offset the apparent cost savings of being able to recruit volunteers. That being said, however, the cost of an NRC study is perhaps somewhat higher than that of a comparable effort carried out by a university or nonprofit think tank and somewhat less than that of a commercial management consulting firm. Under a recent management reorganization of the NRC,[3] efforts are under way to introduce new cost-saving efficiencies, but major savings to carrying out basic studies are not anticipated. In addition, information technology features, such as teleconferencing, Internet communications through email, listservs, and other features, have begun to show efficiency improvements, but such efforts are not likely to fundamentally change the cost of an NRC study, at least in the short term.

Timeliness. The NRC process, which includes selecting and convening a study committee, arranging subsequent meetings for busy people who are serving on a volunteer basis, and navigating a report through peer review, editing, production, and release takes time, sometimes a long time, especially for controversial topics. The recent NRC management reorganization also included some changes designed to decrease the time without affecting the quality, such as streamlining and standardizing lines of decisionmaking authority. Nonetheless, using volunteer experts, addressing controversial subjects, and doing comprehensive work all require time, and the NRC is reluctant to compromise quality for any reason. With an experienced chair and staff and a well-focused task statement, an NRC study committee can deliver reports quickly. For example, when asked to assist the Department of Energy in establishing what ultimately became the Environmental Management Science Program, the study committee produced three reports in 12 months, all of which were peer reviewed. The recent study on climate science requested by the White House was completed in a month, and the recent study assessing the risk of arsenic in drinking water was completed in three months. The dilemma often is that a sense of urgency is expressed for the most controversial studies, which by their nature often take the most time if a comprehensive analysis is required. Less time is usually required if the problem being addressed is more limited in scope.

Committee Authorship Can Also Have Drawbacks. NRC study committees, which are considered an important strength of the NRC, can also sometimes be a weakness. For example, NRC committees are generally made up of distinguished volunteers who have many other responsibilities in their professional lives. Without careful oversight by the committee chair and sometimes NRC management, committee members with the most at stake in a study or perhaps with the most free time available could have a disproportionate influence over a study's deliberations and outcomes. This is why the NRC places such a high priority on strong chairs and professional support staff in managing committees and furthermore takes its procedures for identifying and addressing potential bias and conflicts of interest seriously.[4] Generally, committee members who attempt to abuse their responsibilities as committee members have that opportunity only once and can sometimes be removed while a study is under way.

Committee Makeup To Achieve Consensus. While perhaps more a choice and a trade-off than a weakness, NRC committees are assembled with the intention of achieving a consensus set of conclusions, findings, and recommendations. Consensus, in fact, is part of the charge of most studies, whether requested by agencies or included in congressional mandates. That is, the purpose of the study is to provide consensus recommendations, so that if a very broad set of stakeholders or others are included in the study committee, as one might expect if the purpose is to include all possible perspectives on a problem, a consensus might be difficult or impossible to achieve. Hence, once again without careful oversight of the chair, management, and staff, a committee could, by virtue of its makeup, fail to hear and consider broader perspectives relevant to the study. This is why the NRC places a high priority on the information-gathering phase of a committee's work (workshops, consultants, and other means of information gathering), where such perspectives are heard, studied, and considered and often have a substantial effect on the study's frame of reference.[5] Furthermore, as discussed in Chapter 10, NRC panels tend to produce specific recommendations rather than to lay out and evaluate the pros and cons of a range of alternative policy options, although there is probably no inherent reason why they must do this.

Sponsorship. Most NRC studies are commissioned and paid for by federal agencies through contracts, even those mandated by Congress. This is beneficial in that it helps ensure that what the NRC does is relevant and important. However, on the other hand, it often takes 6–9 months through a government procurement process to initiate an NRC study. In addition, with hundreds of studies ongoing and being planned and delivered, the NRC staff has grown to more than 1,100, and although the work load has been stable or growing for many years, it has also been cyclical in some areas. Hence, careful management oversight is needed to ensure that all staff are used efficiently throughout the NRC, sometimes necessitating temporary assignments to different, albeit likely related study areas within the NRC portfolio. Without such "load leveling," disincentives to objectivity could emerge because all staff are employed "at will" and their continued employment is contingent on contracted studies and is generally a feature of any so-called "soft-money" funded organization. The management reorganization at the NRC included a number of new features for staff development to help guard against the emergence of such disincentives, such as building "core support" for boards overseeing particular study areas. However, the most important safeguard is continued vigilance on the part of management, coupled with the role of volunteer oversight by NRC boards and division committees, to counter some of the disincentives that can emerge from the "soft-money" sponsorship nature of the NRC activities.

Congress Gives Tasks to the NRC

In general, for Congress to ask the National Academies to undertake a study, a bill must be passed, so that the Senate and House must agree and the president must sign the bill in which the assignment is included (although the study request may not appear in the bill itself but, rather, in the report language accompanying the bill). With rare exceptions, assigning the task to the National Academies is not a point of contention, but often it is included in contentious bills. When signed into law, a federal agency must contract with the National Academies for the study. The task statement seeking to implement the intent of Congress is then negotiated between the National Academies and the funding federal agency. Then a negotiation takes place on the amount of funds to be committed by the agency. Sometimes the agency looks at the congressional directive as an unfunded mandate, and negotiations can be difficult. On occasion, Congress may instruct a federal agency to contract with the National Academies for a study on a topic the agency does not want undertaken by an independent group, that is, one that could be critical of the agency. This also can make the negotiations difficult and protracted.[6]

Some Additional Comparisons with the OTA Process

The NRC study process is well developed and often serves an important need of Congress, that is, delivery of an authoritative recommendation on a specific course of action. The OTA, by contrast both by institutional charter and by design of its process, generally did not deliver specific recommendations (Blair 1994). Although not carrying the 100-year-old imprimatur of the National Academies complex, OTA's reports were often also considered authoritative, but OTA's strength was more, as former Chair of the Technology Assessment Board and House Science Committee Chair George Brown (1995) put it, a "defense against the dumb" by elaborating on the context of an issue and informing the debate with careful analysis of the consequences of alternative courses of action without coming to a recommendation of a specific course of action, which often involved value judgements and trade-offs beyond the scope of the OTA analysis.

Both approaches were and can be of great value to congressional deliberation. As an example, in the late 1980s, the OTA was asked to explore the technical issues associated with increasing competition in the nation's electric utility industry. The OTA report (1989) developed and analyzed a series of scenarios for the future structure of the industry, ranging from moving toward a nationally controlled grid to fully deregulating wholesale and retail electricity segments of the industry. The report was detailed and provided a comprehensive background for congressional committees to consider in fashioning legislation regarding restructuring of the utility industry, such as elaborating on the trade-offs associated with each scenario, but it did not select a preferred scenario. It is likely that if the NRC were asked to do such a study, it would be asked to assemble a group of experts, review what is known about the industry and where it is headed, and deliver a specific recommendation on what kind of structure it would recommend and why. The OTA study would likely have been more comprehensive in elaborating on the context of the problem (although not necessarily; some NRC studies include major contracted data collection and analysis efforts) because its resources would have been directed to that end. The NRC's resources would likely have been directed more toward the deliberation of the committee of experts regarding the most desirable course of action, presuming that the assembled experts were already familiar with the problem context. Perhaps too simplistically, the OTA was designed principally to inform the debate, whereas the NRC was designed to help decide on a course of action.

As a different type of example, in the early 1990s both the OTA and the NRC were asked to consider the subject of improving automotive fuel economy and, more specifically, the feasibility of increasing the so-called corporate average fuel economy standards to achieve better fuel efficiency in the nation's auto fleet. The OTA report elaborates on the various trade-offs associated with raising standards and alternative mechanisms for achieving automotive improved fuel economy (OTA 1991). The NRC study (1992), much more specifically, comes to conclusions regarding the technical feasibility of various proposed standards and provides a specific recommendation on a particular set of standards that, in the opinion of the committee, is technically feasible while having minimal or at least acceptable market disruption. The NRC deliverable required that a committee of experts reach a consensus. The OTA study, on the other hand, could seek consensus on facts and analysis (although the process did not require it because the panel of experts was advisory), but it could not come to a specific recommendation on the standards, partly because the agency's charter precluded coming to a specific recommendation in the first place and partly because the advisory panel was assembled with the broadest range of stakeholders and would likely not have been able to reach consensus anyway.

Within the National Academies, some features familiar to the OTA process have evolved over time with NRC studies. For example, the IOM now increasingly hires staff for new studies who are recognized experts themselves in a particular area to work on studies and who consequently take a more active role than was the previous custom in drafting the committee report. This method can increase the already high cost of doing NRC studies, but it has the benefit of increasing the capacity of the study committee to assemble background information efficiently, both as a basis for deliberation and for providing background documentation for the report that would likely not have been included. That is, the report now has more information that can be used both to inform the ultimate decision of the sponsor and to help rationalize the recommendations of the study committee in a more comprehensive manner.

Exploring Mechanisms by Which Congress Might Make More Regular and Systematic Use of the National Academies

The mechanisms for handling congressionally mandated studies are fairly well developed at the NRC, even though many mechanical, contractual, and other administrative procedures are being considered that could improve the timeliness of current NRC efforts, even using the existing mechanisms. In addition, restrictions under the Federal Advisory Committee Act applicable to the NRC's work (see the third part of Appendix 2) as well as the longstanding traditions of the NRC's operations within the National Academies make immediate and sweeping changes difficult.

Nonetheless, on the other hand, one can easily envisage a number of possible additions to the way the NRC does its work that could expand and enhance the role of the National Academies in congressional legislative activities. A number of these possibilities are elaborated on in the following section. Some of these mechanisms already exist but could be used more; some are already under consideration by NRC management; and some would require considerable development to use within the NRC framework.

Existing Mechanisms with the Potential of Increased Use

Direct Responsiveness to Congressional Requests

The NRC responds to Congress indirectly. That is, a study is mandated in legislation and, as noted above, implemented and paid for through appropriations to a federal agency along with accompanying instructions to contract with the NRC to carry out the study. Federal procurement through executive agencies is time-consuming; as noted in an earlier section, the elapsed time from initial contact with a federal agency to a signed contract is generally 6–9 months, even before any work of the study committee has begun. The NRC has been successful in negotiating streamlined "task order" arrangements with some federal agencies that can reduce this time substantially, and increasingly the NRC seeks to create such arrangements. Nonetheless, even then there are complex and sometimes arbitrary procurement rules, at least from the NRC perspective, with which the NRC's responsible study officers, the important professional staff associated with NRC studies, must contend.

Workshops, Roundtables, and Letter Reports

A more or less standard "consensus" study carried out by an NRC study committee takes 12–18 months to complete under customary circumstances. The NRC has developed other mechanisms that take less time but essentially compromise one or more features in the comprehensiveness of the NRC process. Nonetheless, these abbreviated mechanisms can provide considerable value in a shorter time. For example, in recent years the NRC has tried to standardize the definition of different types of convening activities that can assemble groups of experts to discuss topics of mutual interest and to share their expertise. Traditionally these activities take the form of NRC workshops, roundtables, or in some cases continuation of an existing study committee in preparation of follow-up documents to recently released studies known as "letter reports."

Workshops. NRC workshops vary considerably in scale and in the kind of documentation produced. A major distinction that affects the time necessary to convene and document the activity is whether or not the activity is subject to Section 15 of the Federal Advisory Committee Act (FACA) (see Appendix 2). Generally, if the activity results in a statement of the convening body of participants, the statement or report is subject to both standard NRC review procedures and the legal requirements applicable under Section 15 of FACA. Report review and the meeting notice requirements under FACA can add substantially to the time that it takes to complete an activity. If the statement produced by a workshop is either that of an identified rapporteur or is a collection of individually authored statements, the activity's work is considered not to be subject to Section 15 of FACA and can take much less time to complete.

Roundtables. The NRC also convenes roundtables (sometimes known as forums), which in NRC parlance are a type of convening activity that provides a means for representatives of government, industry, and academe to gather periodically solely for the purpose of identifying and discussing issues of mutual concern. NRC roundtable members who are not government officials are appointed to specific terms of membership in the roundtable, and government officials who serve as roundtable members serve ex officio concurrently with their terms in office. An important difference between roundtables and most other NRC study committees is that in the former the dialog is designed to be carried out among a group of individuals acting as individuals and frequently as representatives of particular interests or points of view. That is, in contrast to the institutional requirements for members of NRC study committees, roundtable members are not necessarily selected on the basis of their expertise and are not subject to any institutional restrictions with respect to potential sources of bias or conflict of interest. In general, no public report is issued in conjunction with a roundtable, and no review is required.

Letter Reports. A full-length NRC report providing supporting evidence for study findings is the standard deliverable from an NRC study committee.

Table 8-1. NRC Reports by Division and by Report Type: 2001

NRC division	*Study*	*Meeting summary*[a]	*Letter*	*Total*
Division on Behavioral and Social Sciences and Education	18	11	12	41
Division on Earth and Life Sciences	39	5	13	57
Division on Engineering and Physical Sciences	41	4	17	62
Institute of Medicine Programs	21	13	6	40
Policy and Global Affairs Division	8	6	3	17
Transportation Research Board NRC Program	6	4	15	25
Total	133	43	66	242

[a]Roundtable or workshop.

Source: NRC 2002.

There are occasions, however, when a "letter report" or other abbreviated document may be an appropriate means by which a committee can communicate its views on a subject of particular interest to its sponsor. Letter reports can be planned at the outset for large studies when immediate reporting on findings may be required. Such reports often arise from advisory activities addressing narrowly focused or urgent problems. Also, an NRC study committee may write a letter report when it discovers during the course of its work a critical problem that requires an agency's immediate attention. Finally, an NRC study committee may prepare a letter report as a follow-up activity to release of a major study to provide additional analysis or clarification of the major study's findings and recommendations. Letter reports go through the standard NRC peer review process.

Table 8-1 shows the delivery of different classes of reports by division in 2001.

New Possibilities under Consideration

Enhanced Portfolio of Products

The National Academies are already examining the means by which they deliver information to their sponsors and the public. Greatly enhanced use of Internet-based means of information dissemination, more readable publications, and other means are being considered. We believe that creative means of delivery of information could be very effective on Capitol Hill, improving communications with members, staff, and, ultimately, the public. In 2000 the NRC established an Office of Communications, charged with developing a communications strategy for the National Academies that would focus on enhancing the visibility of NRC work as well as preparing for future dissemination in different formats, especially electronic formats. The strategy (NRC

2001b) developed by the National Academies' Communications Office was endorsed in principle by the NRC governing board in May 2001. The strategy included a prescription of substantial new activity on the part of the NRC program divisions to fulfill the expectations of the strategy.

Clearing House for Expertise Regarding Follow-up Advice

NRC studies are generally event-oriented activities with many carefully tracked milestones, culminating in delivery of a report and sometimes follow-up activities such as testimony before congressional committees or briefings. However, in general, the NRC does not attempt to maintain a reference of expertise to help follow up on a report's conclusions, findings, and recommendations, other than the report itself. Sometimes, volunteers have the time and the energy to be available for such activities, but often they do not. Provision of such a mechanism could give NRC advice a longer shelf life in Congress.

Real-Time Advice

The horrific terrorist events of September 11, 2001, and the subsequent widespread interest in findings ways to contribute to the understanding of the science and technology dimensions of elevated threats of terrorism on the United States have added impetus to provide much quicker turnaround advice to federal agencies dealing with antiterrorism. The NRC is experimenting with convening small groups of experts who then provide advice as individuals, rather than as a group constituting an NRC committee. Such "real-time" advice does not carry the imprimatur of the NRC process, especially the quality control aspects of committee deliberation and peer review, which is why the NRC has traditionally been reluctant to sponsor such activities. It does, however, provide a new means of delivering timely advice.

Longer Term Possibilities That Would Require More Development

Funding Mechanisms

As noted earlier, contracting with federal agencies is a blessing in that it ensures, for the most part, that requested studies are relevant and needed, notwithstanding those requests designed to simply delay a decision, which the NRC attempts to avoid. On the other hand, with contracting come all the complications of federal procurement, as well as the implications of funding uncertainty on strategic planning common in soft-money operations. Further streamlining of contract arrangements with federal agencies could substantially reduce the time necessary to initiate NRC studies.

Direct Contracting with Congress. The NRC could contract directly with Congress or, more likely, with a congressional support agency such as the

General Accounting Office or the Congressional Research Service of the Library of Congress, which would likely reduce the time necessary to negotiate a final contract.

Annual Appropriation. One could make the case for an annual appropriation from Congress, against which NRC studies could be charged. This would require a decisionmaking body for determining which studies could be funded through such a mechanism. In theory, these studies should be able to be initiated much more quickly than studies initiated through the normal mechanism. This body would likely be similar to the Technology Assessment Board (TAB), which performed this function for the OTA. TAB interacted with the Legislative Appropriations Subcommittees in the House and Senate and the relevant authorizing committees for the OTA in the House and Senate as well. Many of the details of such a panel are discussed in Chapter 10 of this book.

There are a number of drawbacks, however, with alternative funding models applied to the NRC. For example, if a central appropriation became large enough to encompass many NRC studies (either directly or indirectly through a congressional support agency), then it might add a new political dimension to the annual work load of the NRC. For studies falling under the annual appropriation consideration, the political "currency" needed for initiating studies is no longer solely based within the committees of Congress, all of which can initiate legislation and funding authority. Appropriations to fund the study, however, would have to come from the congressional appropriations committees, which may or may not act on the funding authority. Rather, the process involves trying to influence a TAB-like decisionmaking body, which is allocating the resources set aside in the appropriation. Moreover, concentration of authority over initiating studies to such a body also concentrates the risk that if a study is viewed as unwelcome by one political faction or another on Capitol Hill, then the annual appropriation can become a convenient means of political retribution that does not now generally exist for the NRC.

For this and a variety of other more internal reasons, the NRC's management has generally discouraged the idea of introducing an annual appropriation from Congress, which is often cited as a possible means for reducing the time needed to initiate an NRC study. It is conceivable that within Congress an equivalent of such an annual appropriation could be set aside within the legislative branch budget and allocated for purposes of initiating "fast-track" NRC studies. However, such a strategy carries many of the same challenges as an annual appropriation, for example, who decides which studies are to be initiated through such a mechanism and how much money is to be allocated for that particular study and, because the NRC is a private institution, what federal procurement vehicle would be used or if a new one would be required.

Indeed, the NRC and the National Academies review the "core and continuing" activities of the NRC annually. If the NRC's governing board were to consider a more direct relationship with Congress, it is likely that such a proposal would provoke extensive discussion with NAS, NAE, and IOM members.

Collaborative Ventures with the NRC

Occasionally, the NRC is able to carry out collaborative ventures with other organizations such as the National Academy of Public Administration and, more recently, the General Accounting Office (GAO). In general, such arrangements are difficult to fashion for a variety of reasons. First, in general, the NRC is reluctant to compromise on its usual extensive independent review process in any activities in which it participates when there is any chance that the resulting product could be confused with a standard NRC committee report. Second, any activity in which the NRC participates in creating a report that includes advice to the federal government is subject to Section 15 of the Federal Advisory Committee Act. These two considerations and others put fairly tight constraints on collaborative ventures with the NRC. That said, one recent collaborative experiment with the GAO may represent evidence of a possible venue for improved technology assessment capabilities for Congress in addition to traditional NRC studies.

The GAO experiment in technology assessment, not necessarily including any NRC involvement, was first proposed by Senator Jeff Bingaman (Democrat, New Mexico) to explore ways, in the wake of OTA's demise, to reconstruct a technology assessment capability within the legislative branch without the major investment of a new organization. In the course of deliberation over the FY2002 Legislative Branch Appropriations Bill, the House–Senate conference report accompanying that legislation instructed the comptroller general to obligate $500,000 to initiate "a pilot program as determined by the Senate and submit a report on the pilot program not later than June 15, 2002."[7] Subsequently, in December 2001, a letter from four U.S. senators[8] articulated the assessment topic for the pilot program as "The Future of U.S. Border Control—The Role of Technology." This request, along with expressed interest in tracking the experiment, was also endorsed by six House members. Recognizing the time constraints, the broad scope of the request, and the relative unfamiliarity of the GAO staff in carrying out a technology assessment, as opposed to the more traditional government audits carried out by the agency, GAO and congressional staff narrowed the scope of the pilot study to the role of biometric technologies in border control. In addition, GAO staff recognized early the need for external technical and policy expertise on biometric technologies and sought help from the NRC to locate and convene groups of experts in these areas. As a result, the NRC organized two workshops in support of this effort, one reviewing the technical alternatives for biometric identification and the other focusing on the range of policy issues associated with deployment of such technologies. Although these workshops did not result in NRC reports, they did provide valuable background information for the GAO study team.

The GAO report on biometric technologies was released in November 2002 (GAO 2002). Although this experiment was difficult to initiate for a variety of reasons and many shortcomings remain in the approach adopted by the GAO in carrying out its first attempt at a technology assessment (see Appendix 3), the external evaluation cited above suggests that the experiment has been more successful than many anticipated, at least given the constraints. That report articulated a number of significant organizational challenges that the external review group felt were necessary to improve the GAO approach, which could then possibly evolve to a more mature technology assessment capability within the legislative branch. Whether the GAO is capable of such reforms remains to be seen, but it seems fair to conclude that the initial GAO experiment has yielded evidence sufficient to at least continue the experiment. The NRC role appears to have been one of the successful features of this approach and may constitute a way in which the National Academies can contribute to a renewed technology assessment capability within the legislative branch, in addition to its more traditional response to congressionally mandated requests for assistance. It provides GAO a degree of access to the National Academies' formidable network of technical expertise.

Conclusions

The NRC has had a longstanding and effective, if at times complicated, working relationship with Congress on even the most contentious issues. Indeed, there are many examples of how Congress has responded directly to the recommendations given in NRC studies. Nonetheless, there are many characteristics of that relationship that could be improved, both to perform the traditional NRC role more effectively and to provide some opportunities to expand that role. However, even with the current arrangement, the role of the NRC in providing advice to Congress will likely gradually increase as the nature of issues facing Congress includes more science and technology related dimensions. Even today, the science and technology features of many important issues facing Congress—energy, environment, health, defense, telecommunications, infrastructure, housing, work force issues, and trade—are becoming more complex and important.

One could envisage a more OTA-like function being carried out within the NRC, but it would involve resolving many internal and external institutional control issues, especially if Congress sought to fund it with an annual appropriation. What would be the decision rule for commissioning an OTA-like study versus a more traditional NRC study? Could the NRC say no, as it does with some kinds of federal agency requests? Who would decide on the allocation of resources from an annual appropriation? With an annual appropriation likely would come an oversight responsibility and a budget authorizing committee jurisdiction, resulting in the complex and sometimes paralyzing (and ultimately fatal for OTA) situation that the OTA faced with multiple committees of jurisdiction and no real champion, at least for that particular function within the NRC. Would it be appropriate for the NRC, as a nongovernmental institution (as compared with the OTA, an agency of Congress), to be so intimately connected to such congressional processes? These are questions, not answers, but they are hard questions.

Acknowledgements

The authors wish to acknowledge the important contributions to the preparation of this chapter of William Colglazier, the NRC's executive officer; James Jensen, director of the National Academies' Office of Congressional Affairs; and Richard Rowberg, associate executive director for communications of the NRC's Division on Engineering and Physical Sciences. In addition, the editorial suggestions of Susan Turner-Lowe, former director of the National Academies' Office of News and Public Information, made this chapter much more readable. However, the blame for any errors rests with the authors.

Notes

[1]As of 2001, the NAS had 2,283 members, of whom 325 were foreign associates; the NAE had 2,207 members, of whom 154 were foreign associates; and the IOM had 1,373 members, of whom 56 were foreign associates.

[2]Each year, NRC and IOM together have almost 600 active committees, most of which are appointed to carry out studies commissioned by federal departments and agencies. These studies involve a level of effort of $130 million in annual expenditures, now distributed fairly evenly across six divisions: Behavioral and Social Sciences and Education; Earth and Life Sciences; Engineering and Physical Sciences; Institute of Medicine; Policy and Global Affairs; and the Transportation Research Board.

[3]The NRC commissioned a Task Force on NRC Goals and Operations in the mid-1990s that concluded with the delivery of a report (NRC 2000), which was delivered to the NRC Governing Board in August 2000. Subsequently, the governing board approved a number of major structural and process reforms that took effect in January 2001.

[4]By comparison and contrast, the OTA process used an authoritative committee of volunteers as an advisory panel or committee rather than assuming authorship of the study itself. It was left to the professional staff, who were responsible for the report, often along with the help of the committee chair, to moderate the roles of advisory panel members. This method permitted easier regulation of the role of the committee, especially if it was not performing as well as expected in its deliberations, but such a practice also sacrificed the authoritativeness of the volunteer experts as authors of the report, an important feature of the NRC process.

[5]By comparison and contrast, in the OTA process, because the advisory panel was advisory and not the authors, the necessity of reaching a consensus was seldom an issue. Indeed, the panel was created to try to collect views of all important stakeholders, rather than to try to achieve a consensus. Perhaps more significantly, however, the OTA seldom tried to issue recommendations. Rather, the project teams sought to analyze the consequences of a particular course of action and elaborate on the context of a problem, often without coming to specific recommendations on a course of action. This practice did not always succeed, but when it did, it provided another source of relief for the study team for having to find consensus among what could be a diverse group with points of view that prevented consensus on many controversial issues. If required to come to a consensus set of recommendations, the OTA model would likely be unworkable for controversial subjects with many opposing points of view.

[6]As a point of reference for the current levels of congressionally mandated activity at the NRC, during the time noted above (102nd–106th Congresses) in which there were

186 congressionally mandated studies, the National Academies conducted approximately five times as many consensus studies directly for federal agencies. The NRC also on occasion carries out studies for state agencies or even private foundations. In a few cases, studies have been self-initiated by the NAS, NAE, or IOM using their limited endowment income on topics these institutions believed important when funding was not available from the federal agencies or foundations.

[7]A detailed examination of the process of the GAO experiment is provided in Fri et al. (2002), a copy of which is reproduced in Appendix 3.

[8]Letter to David M. Walker, Comptroller General, General Accounting Office, signed by Senators Jeff Bingaman, Pat Roberts, Joseph Lieberman, and Pete Domenici, December 17, 2001.

References

Blair, Peter D. 1994. Technology Assessment: Current Trends and the Myth of a Formula. First Meeting of the International Association of Technology Assessment and Forecasting Institutions, Bergen, Norway, May 2, 1994, in OTA 1996.

Brown, George E., Jr. 1995. OTA, Congress's Defense against the Dumb, Closed Down; Congress Left Defenseless. *Congressional Record*. Extension of Remarks, September 29.

Fri, Robert W., M. Granger Morgan, and William A. Stiles Jr. 2002. An External Evaluation of the GAO's Assessment of Technologies for Border Control. Report to the General Accounting Office, October 18.

General Accounting Office (GAO). 2002. *Technology Assessment: Using Biometrics for Border Security.* GAO-03-174. Washington, DC: General Accounting Office.

National Academies. 2000. *Roles of the Committee Chair.* Washington, DC: National Academy Press.

———. 2002. *Survival Guide for Study Directors.* Washington, DC: National Academy Press.

National Research Council (NRC). 1992. *Automotive Fuel Economy: How Far Can We Go?* Washington, DC: National Academy Press.

———. 2000. NRC in the 21st Century: Report of the Task Force on NRC Goals and Operations. Report to the NRC Governing Board, March.

———. 2001a. *Disposition of High-Level Waste and Spent Nuclear Fuel.* Washington, DC: National Academy Press.

———. 2001b. Communications at the National Academies. Report to the NRC Governing Board, May 9.

———. 2002. The National Academies Report Review Database, January.

Office of Technology Assessment (OTA). 1989. *Electric Power Wheeling and Dealing: Technological Considerations for Increasing Competition.* OTA–E–409, NTIS Order No. PB89–232748. Washington, DC: U.S. Government Printing Office, May.

———. 1991. *Improving Automobile Fuel Economy: New Standards, New Approaches.* OTA–E–504, NTIS Order No. PB92–115989. Washington, DC: U.S. Government Printing Office, October.

———. 1996. OTA Legacy on CD-ROM, GPO 052–003–01457–2, available at www.wws.princeton.edu/~ota (accessed May 3, 2003).

9

Expanding the Role of the Congressional Science and Engineering Fellowship Program

Albert H. Teich and Stephen J. Lita

> A certain representative from California was questioning a witness from the U.S. Department of Energy at a hearing held by the House Energy and Commerce Committee. The topic was the Northwest Power bill, and the witness was trying to politely dismiss a proposal for electrical power transfer between the Northwest and California. "We'd like to do that, Congressman, but first we'd have to repeal the second law of thermodynamics," the witness said. As the hearing ended, the Congressman turned to an aide and asked whether an amendment repealing the second law of thermodynamics could be drafted in time for the next hearing. (Schmid 1994)

The 1995 closing of the congressional Office of Technology Assessment (OTA) significantly diminished access to scientific information within the U.S. Congress. Other chapters in this volume have discussed new mechanisms for filling this void, such as the revival of a "new OTA"; modifying or expanding the roles of the General Accounting Office, Congressional Research Service, and Congressional Budget Office; or making greater use of the National Research Council of the National Academies complex. This chapter examines mechanisms that build on an existing, innovative program that has provided science and technology advice to the U.S. Congress for almost 30 years: the Congressional Science and Engineering (S&E) Fellowship Program.

The Congressional S&E Fellowship Program is a collective effort supported by the American Association for the Advancement of Science (AAAS) and about 35 other professional and scientific societies. Each society selects and funds one or more fellows. AAAS runs an umbrella program for all the fellows, including an orientation, assistance with placement, and frequent seminars and social activities. The fellowship program serves several purposes. It is a means by which the selected scientists and engineers can gain first-hand experience in the operations of congressional offices and committees and develop an understanding of the policymaking process. It also brings strong science into congressional decisionmaking by providing highly trained professionals as resources for congressional offices. In addition to possessing expertise within their specialized fields, fellows bring with them an understanding of the importance of the scientific method and the ability to search for and quickly obtain required data. Finally, the fellows become a conduit to the greater science community for the policy lessons learned. Fellows serving on the Hill thus play an important role in helping members, committees, and staff identify the information they need, synthesize it, and interpret the relevant information to policymakers and the general public alike.

Since the formation of the Congressional S&E Fellowship Program in September 1973, more than 760 scientists and engineers have served on Capitol Hill. The program began with 6 fellows. Today the Congressional S&E Fellowship Program brings about 35 fellows to Washington each year. Historically, about two-thirds of congressional S&E fellows have worked for an individual member of Congress, with the remainder working on committee staffs.[1]

In 1994, Anthony Fainberg, a former congressional S&E fellow, collected and edited almost 60 essays penned by former fellows who served on Capitol Hill about their year-long experiences. The volume, *From the Lab to the Hill: Essays Celebrating 20 Years of Congressional Science and Engineering Fellows,* not only captures the effect that the experience has had on the personal lives of the fellows, but also provides concrete examples of how the fellows have partially filled the Hill's science and technology knowledge vacuum and improved related policies. Fainberg (1994) writes,

> The Fellows have changed the political landscape in Washington in a significant way, one that may not be apparent to the general public but that is present nevertheless: Fellows greatly reduced (although not entirely eliminated) scientific illiteracy in the halls of Congress. They have also helped convince Congress of the importance of science and engineering to the country. This broad effect of the Program, although hard to quantify, inevitably has had a major influence on many political decisions that have had a scientific or technological impact over the past two decades. I think the nation is better off than it otherwise would have been.

Many fellows have not been merely a source of temporary expertise but have turned their fellowship years into successful careers on the Hill. The fellowship program counts a number of legislative directors and legislative assistants, on personal and committee staffs alike, among its alumni. It was a watershed moment for the program when 1982–1983 American Physical Society Congressional Fellow, Rush Holt (Democrat, New Jersey), was elected to Congress in 1998.

A Brief History

The germ for the Congressional S&E Fellowship Program grew out of a series of Stanford Workshops on Political and Social Issues (SWOPSI) held at Stanford University during the 1969–1970 academic year. Two physicists, Frank von Hippel and Joel R. Primack, headed a task force in one SWOPSI workshop concerned with congressional decisionmaking and the function that scientists and engineers perform in advising government. According to Jeffrey Stine's history of the fellowship program (1994), the task force based its recommendations on a questionnaire sent to every member of Congress. In the preface to their report, von Hippel and Primack discussed the roles of scientists who have previously provided technical advice to government,

> Their efforts are mainly dedicated to the service of the executive branch of the federal government, and almost all of their advice is secret. We are concerned that the confidentiality of this advice may undermine essential democratic controls on the direction of technology, while at the same time weakening the force of the technical advice itself by depriving it of a visible constituency. We find it difficult to escape the conclusion that the "insider" approach of the "pros" who comprise the advisory establishment of the executive branch needs to be complemented by a more open advisory system for Congress and the public. The technical community has a great responsibility to educate and lead the nation on the constructive development and application of technology. (Stine 1994)

The report came out in support of the creation of a congressional Office of Technology Assessment and also endorsed the fellowship concept. It recommended the creation of a congressional internship program for young scientists and engineers (Stine 1994). Primack and von Hippel wanted these interns to be selected by a national competition, and the winners would spend a year with members or on a committee "with special responsibilities in technical areas" (Stine 1994).

Primack successfully lobbied AAAS to take the lead in the creation of the program, and in October 1971, the AAAS board of directors asked the association's staff to "explore sources of extramural funding for the scientist internship program" (Stine 1994). In early 1972, AAAS established the Congressional S&E Fellowship Program. While AAAS was working on the financial and administrative details of its still inchoate program, the American Society of Mechanical Engineers began its own congressional fellowship program on an experimental basis and placed its first fellow, Barry I. Hyman, in the Senate Committee on Commerce, Science, and Transportation in January 1973.

Buoyed by Hyman's successful start, AAAS Treasurer William T. Golden agreed to anonymously underwrite the costs of placing two AAAS congressional S&E fellows. Empowered with new funding, AAAS created a list of potential fellows and convened a meeting with the American Physical Society and the Institute of Electrical and Electronics Engineers to work out the final details of the program. The three groups agreed to support the nascent program and announced their newly created fellowships in an editorial by Richard Scribner, then director of AAAS's Science and Society Programs, in the April 13, 1973, issue of *Science.* In making his case for the importance of the Congressional S&E Fellowship Program, Scribner (1973) wrote,

> In no way does the active involvement [in the fellowship program] described above mean that the scientific and engineering societies are promulgating the erroneous philosophy that "only science can save the world." However science and technology are crucial elements in the consideration of many problems facing decision-makers.

That first year, 6 fellows were selected. That class performed so well and had such an effect that all 6 were offered congressional staff positions by the end of their fellowship year; 5 accepted (Stine 1994). The fellowship program was continued the next year, and the class of 1974–1975 more than doubled to 14 fellows and added two new sponsoring societies, the American Psychological Association and the American Institute of Aeronautics and Astronautics. In the 1975–1976 year, 16 fellows were placed, and the American Chemical Society and the Federation of American Societies for Experimental Biology were added to the growing list of sponsoring societies.

In his 1973 editorial in *Science,* Scribner drew an important distinction between general informational and analytical resources such as the Congressional Research Service or OTA and the function that the fellows would fulfill:

> The congressional staff includes a few people with strong scientific or engineering backgrounds, but the resources available to congressmen for informing themselves about the technical components of national issues and effectively utilizing existing scientific information are considerably less than those available to the Executive Branch. The reorganized Congressional Research Service, and the emerging Office of Technology Assessment will provide greater informational resources, but the utilization of these by individual congressmen often requires in-office capability. (Scribner 1973)

In his 1994 evaluation of the fellows program, Stine agrees, observing that typically,

> ... congressional members do not need technical information such as the atomic weight of gold or the melting point of iron; they need staff members who are comfortable dealing with science and technology-

> related policy issues. Because of an often unspoken aversion to science and engineering among many staff members, the general scientific and technical literacy of the Fellow [has] proved extremely useful. The Fellows offer ... congressional offices the ability to evaluate, probe, question, and translate technical materials; they also ... [know] where to look for technical information or second opinions. (Stine 1994)

This observation is reflected repeatedly in the commentaries on their experiences by former fellows, which Stine reports, as well as commentaries by a number of more recent fellows that AAAS makes available on their website (fellowships.aaas.org/experiences.shtml, accessed May 6, 2003).

Stine concludes his history of the Congressional Science and Technology Fellowship Program by observing that these programs have

> ... helped expand the disciplinary diversity of legislative staffs not only through the Fellows' temporary presence but also through the Fellows who remain with Congress, and by raising the awareness among congressional members that scientists and engineers can assist their work in meaningful ways. (Stine 1994)

He observes that

> Energy and the environment, economic competitiveness, sustainable development, health care, and debates about big-ticket federal projects are all lending themselves to analysis by technically trained people, and while Congress ... [now employs] ... more scientists and engineers on its permanent staff ... than it did in 1974, their numbers still fall far short of the need at hand, and the Fellows Program helps fill this gap. (Stine 1994)

The majority of fellows work for individual members of Congress. With rare exceptions[2], the rest work on the Democratic or Republican staff of a committee and answer to the ranking Democrat or Republican member on that committee. Therefore, the job of most fellows, like that of all other members of legislative staff in Congress, is to advance a partisan agenda. Because members and committee chairs generally entrust only their own staff with the sensitive tasks of advancing legislation and performing oversight functions, this allows fellows to play a central role in the process. This also means that fellows are frequently not free to move where their expertise is needed. Furthermore, given the breadth of any member's agenda, a solitary scientist will be expected to handle some issues outside of his or her range of scientific expertise.

Fellows generally serve one year in Congress. One year is long enough for a scientist or engineer to gain experience to launch a career in the policy world or to bring this valuable knowledge of Congress back to the scientific community. Either way, fellows and former fellows can bridge the policy and scientific communities. However, this one-year duration also means that fellows cannot provide institutional memory on scientific issues, and by the time they begin to understand the complex ways of Congress, they are ready to leave. Thus, transient fellows cannot serve the same purpose as a stable and permanent organization that understands what Congress needs and what science can offer. As valuable as it is for a handful of members of Congress to have one trusted scientist on their staffs, this is certainly no substitute for a cohesive group of experts who produce nonpartisan reports and briefings intended for all members of Congress.

In addition to its ongoing success on Capitol Hill, the Congressional S&E Fellowship Program has been the model for nine other science and technology policy fellowship programs that are managed by AAAS. These programs provide the opportunity for accomplished and societally aware postdoctoral to midcareer scientists and engineers to participate in and contribute to the public policymaking process of the federal government through placement in various executive branch agencies. These programs have brought another 650 fellows to Washington.

Expanding the Program

By September 2003, the Congressional S&E Fellowship Program will have celebrated its 30th anniversary and placed its 800th fellow on Capitol Hill. Thirty years and 800 alumni constitute a testament to the success of the program, and there are calls for the fellowship program and sponsoring societies to do more to improve the analytical support for fellows—and for Congress—especially in the aftermath of OTA's dissolution. In response, we have identified several possible ways that the program can be modified to improve and expand the analytical support given to congressional science fellows as well as to improve the use of fellows within Congress without compromising the success that the program has enjoyed.

Fellowship Orientation

The fellows bring essential scientific and technical knowledge to the congressional offices and committee staffs in which they work. But to usefully apply that knowledge, they first must gain an understanding of the unique culture and needs of Congress. To this end, each September AAAS hosts an intensive orientation program that features more than 150 speakers over a two-week period for the incoming class of "AAAS Science and Technology Policy Fellows," which is the overall designation that AAAS uses for all 10 of its policy fellowship programs. The class includes the 35 congressional S&E fellows and the approximately 60 fellows serving in executive branch agencies. The orientation serves as an introduction to Washington, D.C., and provides the incoming class with a broad overview of how government works, including the budget process, ethics regulations, and the roles played by Congress and the various federal agencies involved with science and technology issues. On some days, the orientation program is broken into three programmatic tracks: legislative, international, and executive branch. The legislative track designed for congressional fellows offers sessions on the Congressional Research Service, current science and technology issues in Congress, and an insider's view on how a congressional office operates.

In spite of the significant substance imparted during orientation, fellows often talk about the steep learning curve that they face when they begin their work on the Hill. It often takes a month or more before a fellow feels sufficiently knowledgeable to make concrete contributions. To strengthen the learning process, the AAAS fellowship staff has considered extending the two-week orientation to allow the addition of several days of workshops to give the congressional fellows a deeper understanding of the policymaking process, legislative calendar, floor procedure, and the various chamber rules. This extension would delay their placement somewhat, but the costs would be offset by the benefits of additional preparation for their roles.

Improving the Use of Alumni Networks

The 1,500 former AAAS fellows possess a wealth of experience and information about the federal government and science and technology policymaking. This group includes both congressional and executive branch fellows from the various AAAS programs. They are eager to stay connected to their Washington experience, which most describe as one of the high points in their careers.

AAAS maintains a master database of the 1,500 alumni who have passed through the fellowship program. It contains up-to-date information on their positions and activities, as well as information on their previous policy experience and their technical expertise. Former fellows are often called on as speakers during orientation, to serve on selection committees, or to help with the review of fellowship applications. However, an innovation would be to create groups of former fellows who are willing to provide advice, information, or mentoring to the current class. An electronic model would be to create listservs based on specific areas of policy interest such as biotechnology, telecommunications, or energy policy that could provide current fellows with additional resources to rapidly locate experts and information during their tenure on the Hill.

Improving Interaction with the Executive Branch Fellowship Programs

The closure of OTA has forced Congress and the fellows on the Hill to rely increasingly on the executive branch agencies for scientific and technical information. The nine other policy fellowship programs that AAAS administers place scientists and engineers in a dozen federal agencies, including the National Science Foundation, the U.S. Environmental Protection Agency, the Department of State, the Department of Defense, the Agency for International Development, and the National Institutes of Health. Because of the role that the executive branch fellows play in the planning, development, and oversight of agency programs, they are a significant resource for congressional S&E fellows. It is a two-way street—the congressional fellows also provide perspectives to executive branch fellows that are useful in their work. Strengthening the interaction among the various programs through special topical workshops and seminars would help create networks and opens lines of communication among the groups.

Spreading the Wealth

Incoming congressional fellows are not assigned to congressional offices when they arrive in Washington; they spend two to three weeks searching for an office where there will be a good fit. Placement is a personal decision based on interviews and interaction with members or staff, policy interests, party affiliation, and partially on "word-of-mouth" from the experiences of former fellows. For the most part, fellows look for positions where they will be treated as legislative assistants on the permanent professional staff and given responsibilities that will be challenging and potentially have an effect on policy.

Although there have been almost 800 congressional S&E fellows since 1973, fewer than 120 representatives and fewer than 70 senators have placed fellows in their personal offices. In fact, just 27 representatives and 15 senators account for more than half of all congressional S&E fellows serving in their respective chambers. Not surprisingly, clustering has also occurred within certain committees and subcommittees, as fellows gravitate toward those that have a jurisdiction over scientific and technical subjects, such as energy, environment, communications, transportation, agriculture, and health. Similarly, fellows are often drawn to serve on the personal staff of members who serve on these scientifically oriented committees.

The fact that several members and chiefs of staff actively recruit fellows for their offices shows an admiration for the skill sets that fellows bring with them to the Hill, but it unfortunately limits the dissemination of their expertise. These offices and committees were already scientifically literate. This fact may have made the fellows more comfortable and eased the transition to their new roles, but it has narrowed the legislative body's interaction with this valuable science and technology resource and limited the effect that the fellows could have made on a broad range of policies.

To spread the wealth of scientific expertise to new offices and committees, the fellowship program staff could actively market the program to those offices and committees that have not previously hosted fellows. This would not only open new avenues to fellows that were previously unavailable, but will also further interaction between policymakers and the science and technology community. It may also broaden awareness of science and technology policy issues among members of Congress.

The fellowship program staff and sponsoring societies could additionally make an effort to explain to members and their staffs ways that the expertise of the fellows can be used better. Better communication between congressional staffs and the fellowship program would undoubtedly be beneficial to both parties, but this linkage comes with a caveat: the science and technology community cannot expect Congress to change to meet its whims. The burden of offering a desired product falls on the fellowship program, sponsoring societies, and the fellows themselves.

The incoming fellows should be made aware of the important need that Congress has for science and technology advice outside of the science-related committees and encouraged to seek placement in offices that have not traditionally used fellows. This broadening of horizons could be achieved through an orientation session on the legislative track that discusses the new responsibility that the fellows have acquired in the absence of OTA and encourages them to look beyond a comfortable placement.

Expanded Analytical Support

Most fellows report that their time is entirely consumed by the day-to-day needs of the committees and personal staffs on which they work. They may occasionally identify specific modest pieces of technically focused policy analysis that could assist the Congress in its work, but they typically have neither the time nor resources to perform or commission such analysis.

Whereas the scale of any fellow-based analysis effort would necessarily have to be modest, if resources were available, fellows might usefully draw on a network of academic research units and nonpartisan think tanks to seek modest analytical support. On occasion, fellows with policy–analytic experience might also work together on some analytical efforts.

Resources for such undertakings could be provided by congressional sources. Alternatively, one or more private foundations or wealthy individuals might establish an endowment fund at one or more universities or think tanks, or with AAAS, to support such undertakings. Of course, were such a fund to be created at AAAS, managing how it was allocated would raise, on a smaller scale, all the concerns about balance and bipartisan oversight that are discussed in other chapters of this book.

Conclusion

The closure of the congressional Office of Technology Assessment has forced Congress to look outside of itself for balanced and neutral science and technology analysis and synthesis. Organizations such as the General Accounting Office, Congressional Budget Office, and Congressional Research Service have been suggested as internal mechanisms that can partially fill the void left by the loss of OTA. Suggested external sources of analysis include the National Academies complex, policy think tanks, and professional and scientific societies. Of these external mechanisms, one has been providing science and technology analysis on the Hill for almost 30 years: the Congressional Science and Engineering Fellowship Program.

Fainberg (1994) wrote, "Most Fellows came to Washington not just to improve their own career prospects (and, indeed, that was not at all the reason why some participated); but they also came with the altruistic intent to make a positive impact on the nation, the scientific community, and the relationship between the two." In truth, many congressional S&E fellows have made an impact during their fellowship year, and many have stayed in Washington and continue to influence science and technology policy. By almost any standard, the Congressional S&E Fellowship Program has been a success, benefiting the fellows, the offices and committees they served, and the nation at large. Nonetheless, there is always room for improvement.

We have identified several ways to expand the analytical support given to congressional S&E fellows and thus improve the use of fellows within Congress and to strengthen the program as it enters its fourth decade. Possible changes include expanding the orientation program to provide a deeper introduction to congressional procedures and the policymaking process; expanding the use of the alumni network to take advantage of the experiences of former fellows; increasing the interaction among the congressional and executive branch fellows; and encouraging each new fellowship class to seek placement in offices and committees that have not previously hosted a fellow. These are not revolutionary proposals. They are incremental changes that, we hope, can help to make a fine program even better.

Acknowledgement

We are grateful to Claudia J. Sturges, director of the AAAS Science and Technology Policy Fellowship Programs at AAAS, for her advice and assistance in the preparation of this chapter.

Notes

[1]From 1973 to 2001, 230 congressional fellows served time in the offices of representatives, 236 in the offices of senators, and 230 have worked on the staffs of various committees. (Several fellows had multiple assignments during their fellowships, and many members and committees took multiple fellows during a given year.)

[2]A few fellows have worked for congressional support agencies such as the Congressional Research Service, serving in roughly the same capacity as permanent staff of those agencies.

References

Fainberg, Anthony (ed.). 1994. *From the Lab to the Hill: Essays Celebrating 20 Years of Congressional Science and Engineering Fellows.* Washington, DC: American Association for the Advancement of Science.

Schmid, Charles. 1994. Preface. In *From the Lab to the Hill: Essays Celebrating 20 Years of Congressional Science and Engineering Fellows*, edited by Anthony Fainberg. Washington, DC: American Association for the Advancement of Science.

Scribner, Richard A. 1973. Scientist Congressional Fellow, editorial. *Science* 180: 139. April 13.

Stine, Jeffrey K. 1994. *Twenty Years of Science in the Public Interest: A History of the Congressional Science and Engineering Fellowship Program*. Washington, DC: American Association for the Advancement of Science.

10

A Lean, Distributed Organization To Serve Congress

M. Granger Morgan, Jon M. Peha, and Daniel E. Hastings

This chapter explores the pros and cons of creating a small, permanent institution within the U.S. Congress that would receive requests for studies from committees and then farm them out to one or more previously approved nongovernment, nonprofit organizations that have agreed to conduct studies in accordance with a set of procedures designed to ensure balance, neutrality, and completeness. Such a model was proposed in 2002, albeit in a less elaborated form, in language calling for the establishment of a National Technology Assessment Service in the Senate version of the energy bill (Senate Bill 517).

Such a model would be new in a U.S. context, but Chapter 5 explains that several of the assessment institutions that serve European parliaments follow a somewhat similar distributed model. Likewise, if the General Accounting Office continues to experiment with technology assessment activities, some of these will probably be undertaken using elements of such a model. However, in the discussion that follows, we limit our discussion to consideration of a small, free-standing unit housed within Congress.

Advantages of this model include (1) the fact that it would support a wider range of expertise than can ever be assembled in an organization that is housed entirely within the legislative branch and (2) the fact that it minimizes the additional staff and administrative infrastructure required to support technology-focused analytical capabilities within the legislative branch.

The Small Permanent Institution within Congress

Under this model, a modest number of previously approved analysis groups in organizations such as think tanks and universities would stand ready to perform analysis on request by a bipartisan, bicameral committee of the House and Senate. Administration of these requests would be handled by a small, permanent committee staff.

The committee would have a staff of 8–12 technical professionals, most of whom would be expected to have a Ph.D. (or other relevant advanced degree) in fields such as science, engineering, medicine, mathematics, statistics, and quantitative policy analysis.

Requests for major studies could come only from committees of Congress (although, as discussed below, an arrangement might be made to also support modest requests from individual members for specific technical advice). Priority would be given to requests that come on a bipartisan, bicameral basis (i.e., from both the ranking majority and minority members of one or more committees in both chambers).

The permanent staff would be responsible for refining such requests, and sometimes consolidating multiple requests, and for making arrangements for an outside organization (or organizations) to perform a study. The joint committee, based on recommendations of staff, would make final approval of all requests for studies and of all final study reports.

Selection, Approval, and Oversight of Outside Organizations

How would the eligible analysis groups be chosen? First, there would be some basic rules for eligibility. For example, one might require that the groups be well-established nonprofit, nonpartisan organizations with some minimum staff size; range of available expertise; record of interdisciplinary, policy-focused research and analysis; and professional productivity and publication. Organizations interested in being considered could be required to prepare a proposal outlining their qualifications and areas of expertise, including resumes of potential participants. Applicants would also be asked to provide cost information for studies of two or three standard sizes to be performed on specified time scales. For example, proposals might be required to provide prices for studies of the sizes in Table 10-1. Pricing for studies should be standardized across all organizations (e.g., perhaps the average of the three lowest cost proposals from approved organizations).

Each year a small number of organizations might be approved for participation for a fixed term of a few years. In that way, a few new organizations might be added each year and a few might rotate off unless they write successful renewal proposals. Selection could be done by the committee staff (based perhaps on both the proposal and a site visit), with final approval of staff recommendations by the bipartisan, bicameral committee. A standing contractual arrangement would be put in place with each approved organization so that there would be no contracting delay when a specific study was commissioned. Most funds should go to pay for actual studies; the cost of the base contract could be nominal, e.g., $1 per year.

Table 10-1. Study Durations and Person-Years to Completion

Study duration (months)	*Study size (person-years)*
3	0.5
3	2
6	2
6	4
9	4
9	6
12	4
12	6

When a particular request for a study came in, the staff would select one or more approved organizations that appear to have relevant qualifications and hold discussions with them to determine their interest and ability to respond on the topic, given the time and resource levels available. In some cases, either to achieve greater depth on an important issue, or to achieve multiple perspectives on the same set of issues, it might be desirable to draw on expertise from more than one organization or for approved organizations to involve other experts as consultants or subcontractors. As the direct contractor, the approved organization would be held accountable for the subcontractors that failed to meet accepted standards of quality, timeliness, or neutrality. It is probably not desirable to allow new organizations to be approved on an ad hoc basis outside of the annual solicitation because this could give rise to various abuses.

Doing analysis for Congress is different from doing analysis in other environments. Because of the highly political nature of the congressional environment, if analysis is to be widely accepted, it must avoid making value judgements or offering policy prescriptions. Analysis can usefully serve the broad needs of Congress by doing the following:

- framing problems (i.e., helping members and their staff understand how to think about an issue);
- identifying topics that are and are not important so that time is not wasted on irrelevancies or wrong ideas (i.e., playing a role similar to that of "stipulations" in a lawsuit); and
- identifying the important policy choices that Congress must make.

Two strategies have been particularly helpful in achieving these objectives in the past. First, an external advisory committee that represents a broad range of interests and constituencies can be used to provide advice and guidance to those performing the study. Although such a highly diverse committee would typically be unable to reach consensus, and thus should not be responsible for approving the study, the very different technical and political perspectives that its members can bring to bear on the problem can help the staff to avoid imbalance and consider the many arguments and perspectives that various stakeholders bring to the topic. Advice should be sought from such a committee in the early stages of a study and again as the study nears completion. When speed is important, it might be possible to hold these meetings electronically. A second strategy, which has proven effective in the past, is to suggest conclusions conditional on a range of possible congressional policy preferences. Thus, a report may contain a range of statements of the form "If Congress wishes to <achieve *x*>, then it should <do *y*>." In this way, a variety of policy options can be discussed without making politically controversial recommendations. An example is reproduced in Box 10-1.

Typical think tanks and policy-oriented academic departments are not used to using such strategies in the analyses that they do. For a distributed system to produce effective products, it would be necessary for approved groups to have participated in a process of training and sensitization so that they would be able to produce analysis that meets the special needs of Congress and not make recommendations that place them on one side or the other in heated policy debates.

Study reports should consist of three elements: a one-page summary for members written in clear lay language, a slightly more detailed and slightly more technical summary of no more than five pages that provides a more detailed review, and a full report written in semitechnical language that does not exceed 50 pages. Additional background material and supporting documentation could be provided over the Internet. Again, many think tanks and policy-oriented academic departments do not have experience in producing concise nontechnical and semitechnical documents. Part of the criteria for selection of approved groups should be a demonstration of this ability.

One or more members of the permanent staff of the bipartisan, bicameral committee would be assigned as liaisons to each commissioned study and would have responsibility for ensuring that the study is proceeding on schedule; that it is meeting the requirements of balance, neutrality, and completeness; and that the pieces are fitting together in a way that will meet congressional needs.

Figure 10-1 illustrates the life cycle of a typical study under this model.

Budget

Studies could be paid for in several ways. The bipartisan, bicameral committee would receive an operating budget sufficient to cover the costs of its staff and to fund at least a small number of studies each year. In addition, individual committees (or groups of committees) could fund or co-fund studies from their operating budgets, although it seems unlikely that that option would be exercised frequently. Third, in much the way that Congress currently mandates executive branch agencies to commission studies through the National Research Council, it could mandate that executive branch agencies commission studies through the congressional assessment process, so that in the

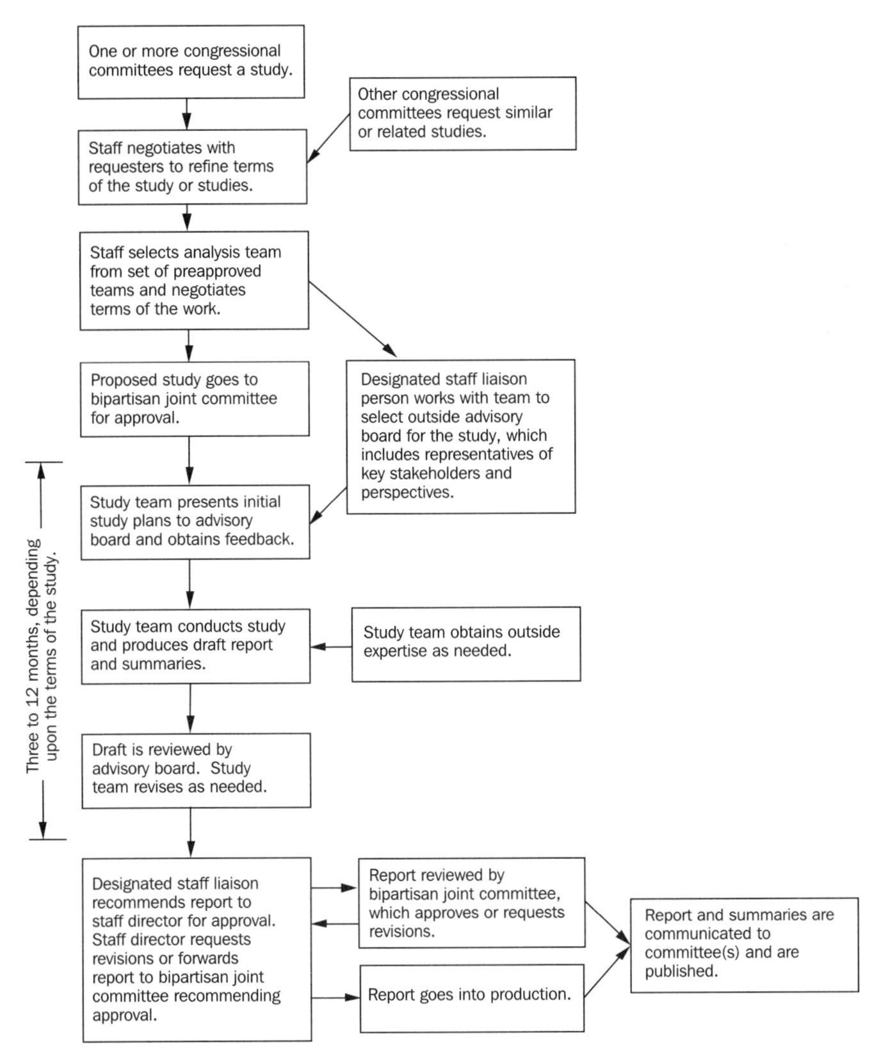

Figure 10-1. Life Cycle of a Typical Study Performed by the Proposed Lean, Distributed Organization

future, congressional committees will have the knowledge and insight that they need to continue to perform their responsibilities in an informed manner.

Finally, Congress could appropriate funding for a specific earmarked study. This only partially solves the funding problem because the appropriations committee cannot easily fund a study without removing funding from something else. In addition, it carries a serious danger. The appropriations committee could in effect replace the bipartisan, bicameral committee. Also, selection of studies would become subject to logrolling, wherein members of Congress support each others' favorite projects, and all the frantic last-minute uncertainties and machinations of the appropriations cycle.

Box 10-1
Congressional Alternatives

Note: This is an example of the "If Congress wishes to <achieve x>*, then it should <do* y>*" format drawn from the OTA report titled* Launch Options for the Future: A Buyer's Guide *(OTA 1988).*

Congress could choose to support the development of many different types of space transportation vehicles. To determine which of these alternatives is most appropriate and most cost-effective, Congress must first make some broad decisions about the future of the United States in space. A commitment to key space program goals will entail a similar commitment to one or more launch vehicle systems. Although highly accurate cost estimates do not exist, the analysis in this study suggests that some launch systems are more economical than others to accomplish specific missions.

If Congress wishes to:	*Then it should:*
Limit the future growth of NASA and DoD space programs:	Maintain existing launch systems and limit expenditures on future development options. Current capabilities are adequate to supply both NASA and DoD if the present level of U.S. space activities is maintained or reduced.
Deploy the Space Station by the mid-90s while maintaining an aggressive NASA science program:	Continue funding improvements to the Space Shuttle (e.g., ASRM and/or LRB) and/or begin developing Shuttle-C: The current Space Shuttle can launch the Space Station, but will do so more effectively with improvements or the assistance of a Shuttle-C. Although Shuttle-C may not be as economical as other new cargo vehicles at high launch rates, it is competitive if only a few heavy-lift missions are required each year.
Send humans to Mars or establish a base on the moon:	Commit to the development of a new unpiloted cargo vehicle (Shuttle-C or Transition launch vehicle or ALS) and continue research and funding for Shuttle II and the National Aerospace Plane. A commitment to piloted spaceflight will require a Shuttle replacement shortly after the turn of the century. Large planetary missions will also need a new, more economical, cargo vehicle.
Continue trend of launching heavier communications, navigation, and reconnaissance satellites and/or pursue an aggressive SDI test program to prepare for eventual deployment:	Commit to the development of a new unpiloted cargo vehicle (Transition launch vehicle) by the mid-to-late 1990s. In theory, current launch systems could be expanded to meet future needs; however, new systems are likely to be more reliable and more cost-effective.

Deploy SDI and/or dramatically increase the number and kind of other military space activities:	Commit to the development of a new unpiloted cargo vehicle (Transition Vehicle or Advanced Launch System). Current launch systems are neither sufficiently economical to support SDI deployment nor reliable enough to support a dramatically increased military space program.

Meeting the space transportation needs of specific programs is only part of the reason for making changes to the current launch systems. Congress may wish to fund the development of critical new capabilities or improvements to the "quality" of space transportation, or Congress may wish to ensure that funding serves broader national objectives.

If Congress wishes to:	*Then it should:*
Maintain U.S. leadership in launch system technology:	Increase funding for space transportation basic research, technology development, and applications. Maintaining leadership will require an integrated NASA/DoD technology development program across a range of technologies. Focused technology efforts (ALS, Shuttle II, NASP) must be balanced with basic research.
Improve resilience (ability to recover quickly from failure) of U.S. launch systems:	Fund the development of a new high capacity, high reliability launch vehicle (Transition launch vehicle or ALS) or expand current ground facilities or reduce downtime after failures or improve the reliability of current launch vehicles. At high launch rates, developing a new vehicle is probably most economical.
Increase launch vehicle reliability and safety:	Aggressively fund technologies to provide: 1) improved subsystem reliability; 2) "engine-out" capability for new launch vehicles; 3) on-pad abort and in-flight engine shutdown for escape from piloted vehicles; and 4) redundancy and fault tolerance for critical systems.
Reduce environmental impact of high launch rates:	Limit the use of highly toxic liquid fuels and replace Shuttle and Titan solid rocket boosters with new liquid rocket boosters or clean-burning solid boosters. The United States will be relying on Shuttle and Titan vehicles through the turn of the century. As launch rates increase, the environmental impact of the Shuttle solid rocket motors and the solid and liquid Titan motors will become more important.

Allowing authorizing committees to mandate that an agency fund a study carries a similar risk. In addition, there is the risk that the agencies themselves, and not the congressional client, might craft the wording of the final study request because the agencies would be the direct sponsors. In that case, would we still have an organization that advises Congress and only Congress? Perhaps not.

For these reasons, we believe that by far the best funding arrangement for an analysis activity is a direct appropriation sufficient to cover both staff and study costs as part of the annual legislative appropriations bill. This appears to be the only way to ensure that the bipartisan, bicameral committee will not be bypassed. In addition, as discussed below, there need to be clear rules on how to set priorities to choose among study requests in the event that the number of requests exceeds the allocated funds. We believe that these same arguments also apply to funding of the models outlined in Chapters 11 and 12.

The enabling legislation for the bipartisan, bicameral committee should clearly establish that it has the authority to approve or disapprove all study requests, independent of the source of their funding. If Congress decided it wanted to expand the number of studies beyond what could be supported by direct appropriations, using one or more other funding options, this provision would ensure that it could only happen in concert with and with the approval of the bipartisan, bicameral committee.

Discussion

What are the advantages and limitations of this model? One needs a wide range of expertise to address the large number of issues of science and technology that come before Congress. If one tries to maintain such a range of expertise within an agency of Congress, that necessarily makes for a rather large staff. It is infeasible to keep an expert on staff for a topic that might come up every 5–10 years. Permanent staff will inevitably gain breadth of knowledge and not depth, and depth is important. Think tanks and universities within the United States have a wealth of expertise on which Congress could draw for analytical assistance. Much of this expertise is maintained in part through research support from the National Science Foundation, the National Institutes of Health, the Defense Advanced Research Projects Agency, and other federal funding sources. Most such experts have no easy way to share that expertise with the policymakers who appropriate their funding. This model allows Congress to draw on a range and depth of expertise that could never be achieved with an internal staff.

However, it is also important to involve some permanent staff. Permanent staff working for Congress are very different from outside experts. Permanent staff understand the needs of Congress. They can communicate effectively with their clients, and they are a natural place for institutional memory. Also, it is much easier to ensure objectivity among employees who must answer to a bipartisan congressional committee.

The model proposed here seeks to gain the advantages of both internal and external experts, while minimizing the disadvantages. Because of the permanent staff, studies can be expected that meet the unique needs of a congressional audience. Because they can draw on the best and brightest around the country, they can find all the experts they need in the specific issues of interest to Congress at the time.

The quality and skills of the small permanent staff would be a key factor in the operation of this model. These accomplished professionals must understand the issues of policy analysis, have good taste and high-quality standards that allow them to identify good and bad analysis, and broad experience that allows them to make informed choices among analysis groups. They must be sensitive to the needs and constraints of working for Congress and effectively communicate these through firm and effective guidance to the study teams. They must supply or obtain the important institutional memory of how Congress has dealt with relevant issues in the past. They need powers of persuasion and diplomatic skills so that they can negotiate appropriate study definitions and, when more than one group is involved in a study, ensure harmonious collaboration.

How many professional technical staff would such a model require? At least two will be needed for general coordination: a staff director and an associate director. At least two would also be needed for final technical editing and other issues of product quality control. Individual staff liaisons or study officers could probably supervise and coordinate two or at most three studies at any given time. Thus, with a staff of 8–12 technical professionals it should be possible to manage a set of about 10–20 studies at any given time.

In addition to detailed analysis, Congress often needs rapid access to individual subject-matter experts. For example, a member might want to know, "How hard is it to eavesdrop on a cellular phone call, and is it technically possible to detect when someone is eavesdropping?" This is a sufficiently technical question that the Congressional Research Service (CRS) would likely find it difficult to produce a technically up-to-date answer. Unless they have a special relationship with a technical expert, faced with such a question, a typical staffer is likely to call a cellular company or a privacy advocate, both of whom are likely to give a partisan answer. In this case, the answer can be given in about three pages, it is pretty noncontroversial, and a number of experts could write such a response.

One way to increase the level of interaction between the proposed analysis organization and individual members would be to develop a mechanism to support such requests. There are two advantages: it would increase the ease with which members could acquire such information and it would increase contact between individual members and the analysis organization. Such individual contact has worked to CRS's advantage. Its relatively limited and highly informal service for individual members almost certainly worked to the disadvantage of the Office of Technology Assessment (OTA). There are, however, three problems. First, because budgets are inevitably limited, there is a resource allocation problem. This might be addressed by setting aside a modest budget to serve individual member requests on a first-come, first-served basis and then developing a billing system to handle additional individual requests once that budget has been exhausted. If, over time, requests regularly exceed the budget, pressures might develop among members to grow the level of support. Second, there is some danger that individual requests might draw participating experts into expressing policy preferences and opinions that might undermine their credentials for performing balanced analysis. That problem might be minimized by having all requests flow to experts through the permanent staff and all responses flow back to members through that staff. In that way, a group of politically sensitive reviewers could be on the lookout for potential pitfalls, both in problem framing and in responses. Finally, it would be important not to let this mechanism create competition with the CRS. Thus, as part of the process of vetting questions, permanent staff should routinely check with CRS before approving questions for consideration and should refer less technical questions to CRS when they believe that CRS can handle them.

We have restricted the participating analytical organizations to nonprofit organizations, mentioning specifically universities and think tanks and stipulating that, as stated early in this chapter, the organization should be

> ... well-established nonprofit, nonpartisan organizations with some minimum staff size; range of available expertise; record of interdisciplinary policy-focused research and analysis; and professional productivity and publication.

We impose these limitations in an effort to ensure that the selected groups will achieve our previously stated goal of conducting "studies in accordance with a set of procedures designed to ensure balance, neutrality, and completeness." However, one of the potential difficulties with this model may be the difficulty of achieving this end.

Many nonprofit organizations have decided ideological preferences or orientations. Those with strong orientations should probably be excluded, although developing an acceptable standard for making such selections is likely to prove problematic. Ensuring that the selected organizations will do their best to perform studies that are perceived as balanced and neutral is essential to the success of the model.

Even when a group has come as close as possible to this ideal, there may be temptations in the political environment of Congress to charge bias when the results are politically unpalatable to a particular person or group. On particularly sensitive topics, there may be advantages to involving more than one analysis group. However, we strongly advise perseverance with the objective of striving for balance, neutrality, and completeness within every product. An alternative, more adversarial model, in which groups espouse specific ideological perspectives, would not work in the congressional setting. Analysis from this process should be the input and foundation for subsequent political debate and value judgement, not the context for such judgements.

Table 10-2. Criteria for Establishing Priorities among Study Requests

Priority level	*Criteria*
Highest	Study requested by both the chair and the ranking minority member of one or more committees in both chambers Study requested by both the chair and the ranking minority member of more than one committee of one chamber
Medium	Study requested by both the chair and the ranking minority member of one committee in one chamber Study requested by a number of individual members of both parties from both chambers Study requested by the chair of one or more committees in both chambers Study requested by a number of individual members of both parties from one chamber Study requested by the chair of more than one committee of one chamber Study requested by members of only one party from both chambers
Lowest	Study requested by members of only one party in one chamber

We have noted the importance of establishing clear rules for choosing among study requests in the event, as sometimes happened for the OTA, the volume of requests exceeds the available budget. Clearly the bipartisan, bicameral committee should be able to exercise some judgement about the importance of topics for the business of Congress and should be able to block studies that do not look like they would make effective use of available analytical resources.

We believe that the highest priority should be attached to bipartisan, bicameral study requests that come from multiple committees. We impose these conditions for several reasons. First, there needs to be a mechanism to limit the demands placed on the analysis organization. Second, it is principally committees that need analysis on longer term, more complex issues requiring foresight, analysis, and synthesis of the kind this organization will undertake. Third, in the highly political atmosphere of Congress, no advisory process will survive for long if its products appear to adopt a partisan perspective or principally serve the needs of one side in a partisan debate. Hence, we suggest the criteria presented in Table 10-2 for establishing priorities among study requests.

We have specified that the process be overseen by a bipartisan committee of the two chambers that controls its own small staff. Alternatively, one might contemplate establishing a staff to perform the same functions within an existing congressional agency such as CRS. However, without the parent committee to approve the choice of studies, vet studies as they are produced, and run political interference with members and committees, we believe that the process would be much more politically vulnerable.

This model faces two potentially significant problems. First, it requires the establishment of a new bipartisan, bicameral committee and the allocation of at least some funds that would appear on the congressional budget. This committee would be similar to the former Technology Assessment Board of OTA. Indeed, it might be possible to simply reactivate that board with a new mandate, thus avoiding the need to create a new entity. Although it is hard to see how to avoid at least some budget for the operation, if limiting the size of that budget became critically important, that might be done by some of the strategies outlined in the last section (called Budget). Second, the selection of outside analysis groups faces a risk of becoming geographically politicized. Some geographic distribution among the analysis groups is desirable, but it would be undesirable if the state or congressional district in which groups are located became the predominant consideration in their selection.

Reference

Office of Technology Assessment (OTA). 1988. *Launch Options for the Future: A Buyer's Guide.* July.

11

A Dedicated Organization in Congress

Gerald L. Epstein and Ashton B. Carter

As Bruce L.R. Smith and Jeffrey K. Stine make clear in Chapter 2, Congress is awash in analyses, arguments, advocacy, and advice. Any institution that Congress creates to sift through this sea of information must stand distinctively apart from all the study shops and interest groups that vie for congressional attention. Such an institution must be both *authoritative* and *credible* to warrant a position of unique responsibility and trust. In this chapter we outline a model based on the creation of a dedicated organization within the legislative branch that we believe best satisfies both criteria.

The authority of an institution that can provide balanced and independent scientific and technological advice to Congress on complex, large-scale questions that require foresight, in-depth analysis, and synthesis depends directly on how closely the institution is associated with Congress. Congressional willingness to act on the institution's analysis and advice, and to allow it to address the most important, controversial, or politicized issues, will depend on the strength of this connection. We believe that only a dedicated congressional support agency would be sufficiently linked to Congress, and therefore sufficiently authoritative, to take on such a central advisory role. Studies that matter require an institution *of* the Congress, not just one that works *for* it.

The credibility of such an advisory institution depends on a number of factors. It must be disinterested and independent, with Congress supplying all its funding and serving as its only client. The institution must be free of political bias, both in project selection and in project execution. The studies themselves must be performed competently. Their scientific, technical, and analytic content must be sound, and they must address issues in their appropriate contexts, addressing all relevant interests and points of view. In this chapter, we describe political, technical, and contextual oversight mechanisms for the agency and its projects that should ensure their credibility.

Project Selection, Quality Control, and Objectivity

Ensuring the relevance of agency work to the congressional legislative agenda suggests that the agency should be given assignments by congressional committees, where the substantive work of Congress is largely conducted. The complexity of the topics to be addressed by this agency and the scope and depth of its analysis imply that each study would require a substantial investment. The agency would be able to support only a limited number of ongoing activities, making it impossible to respond to the requests of individual members of Congress. To arbitrate among committee requests, and to protect the agency from capture by partisan political interests or by the particular agendas of a few committees, political oversight would be required. Such oversight would most effectively be provided by a bipartisan, bicameral oversight committee that does not have potentially conflicting legislative responsibilities.

The quality of the agency's analytic product can be no better than that of the data and assistance it receives. Having a direct congressional imprimatur provides the best motivation for possibly reluctant parties, including executive branch agencies, to cooperate fully in the agency's activities. It also attracts the active participation of the nation's leading experts as advisers, reviewers, and contributors—individuals who will be willing to clear their calendars, and even their consulting fee schedules, for the opportunity to help the agency help Congress. Quality control of the agency's plans and products would be provided by expert advisory panels selected for each study. These advisory panels would offer information, advice, and critical review. They would consist both of top substantive experts, who would help ensure the study's scientific, technical, and analytic quality, and of stakeholders representing all the major interests involved in the study, who would help ensure relevance, context, balance, and fairness. To preserve the agency's accountability for its own work and eliminate the need to seek a consensus that would likely be impossible to achieve, the advisory panels would not take responsibility for the final report. However, their views would provide an important measure of the study's validity and freedom from bias.

The authority to approve each study's final release would rest with the congressional oversight committee, which would review, among other things, the comments of that study's advisory panel and other expert reviewers. Given the diverse political views held by members of the oversight committee, and recognizing the extent and level of detail represented in the report text, such approval would not connote the agreement of all committee members with all individual aspects of the study. Rather, it would signify that the study's process was sound, that its findings reflected objective analysis, and that it avoided policy recommendations based on value judgements or political preferences that should be reserved for elected officials.

The ultimate test of a study's quality and fairness would be the reception it gets from the wider community that Congress represents and to which it listens. Studies that raise the level of debate and that successfully explain highly contentious and complicated issues will be of great interest to those outside constituents who are actively engaged in those issues. Moreover, the same attributes that make the agency useful to Congress—its role as a disinterested, authoritative, and technically competent arbiter—should make its studies particularly attractive to outside observers. In their own interactions with Congress on these issues, these outside observers and constituents—as educated by the agency's work—will serve as an alternate means of delivering the agency's work to Congress. Congress does not exist in a vacuum, and a study that is not credible to these wider constituencies should not and will not be credible to Congress. Therefore, contributing to the public discussion and debate on controversial, technology-intensive issues should be a primary objective, and not an ancillary consequence, of the agency's work.

Project Execution, Staff Development, and Institutional Memory

In a dedicated congressional support agency, studies would be conducted by project teams operating within programs defined by issues (e.g., international security and space) or areas of technology (e.g., communications and information technology). Such a program structure would allow the agency to build and retain expertise in areas of high interest to Congress and to develop relationships with members of Congress and congressional staff that span individual studies. Perhaps more importantly, it would support a core of analysts and managers who understood the agency's analytic process, which—given the agency's unique mission and environment—would require a combination of analytic and communications skills and political sensitivities unlike those of any other institution.

At the same time, a significant fraction of the agency's employees should rotate through on shorter term assignments. The agency would not be able to maintain in-house expertise in all technical and policy areas of interest to Congress. Short-term employees, along with consultants and resources for contract support, would provide specialized skills appropriate to specific studies, particularly as new technologies evolve and issues emerge. Moreover, staff renewal would help provide a sense of dynamism and ensure staff quality.

Comparison to Other Advisory Models

Of all the advisory mechanisms discussed in this book, we believe that this model has the most direct connection to Congress. The stronger this connection, the more comfortable Congress will be relying on the institution to address politically sensitive topics; the better Congress will be able to act on the results; the more willing outside experts will be to lend their time, energy, and expertise; and the more distinctive and authoritative the agency's work will be. Unlike the model described in Chapter 10, in which a small central congressional staff allocates studies among a set of prequalified external organizations, this model places the entire analytic organization under direct congressional supervision. In our view, studies farmed out to outside organizations would not have the congressional imprimatur of those conducted by an in-house agency. If their funding were "laundered" through assorted executive agency budgets to avoid showing up in a single, high-visibility legislative appropriation line item, these studies would risk losing their distinctive congressional imprimatur altogether.

The model outlined in Chapter 7, which implements the science and technology advisory function within an existing congressional support agency, shares with this model the advantages of enjoying a direct congressional imprimatur. However, those advantages are not fully realized without also adopting the oversight structure, advisory panels, and study processes described in this model. If this structure and these processes were to be instituted in an autonomous unit of some other congressional support agency, then in effect the Chapter 7 model becomes this chapter's model, but without the explicit recognition that scientific and technological advice is important enough to warrant its own separate agency.

Chapter 8 places this advisory function in the National Academy of Sciences. This plan not only moves it outside Congress, with all the drawbacks already cited, but also gives it to an institution with its own mission and constituency, creating the potential for divided loyalty or conflict of interest. The model articulated in Chapter 12 creates a dedicated institution outside Congress, free of potential conflict of interest so long as it is not permitted to support other clients and is not embedded within any other organization. However, the weaker connection to Congress lessens the institution's distinctiveness and authority and increases the risk of creating just another shop to produce studies.

Measures of Success

The objective of the support agency described here is to be useful to Congress. However, given the complexity of its task and the diverse needs of its client, measures of success are not readily evident. It will be particularly important for the agency to come to agreement with its overseers, requesters, and funders on what constitutes success. In this respect, dispelling unrealistic expectations will be as important as achieving more realistic ones. One such unrealistic expectation, described in Chapter 4 by David H. Guston, is the "silver bullet" model of policy analysis—that some hitherto undiscovered truth will emerge from a study and cut through a Gordian policy knot. Topics addressed by the agency are likely to have been around for some time, and it is unlikely that a new study will produce some blinding insight that dramatically affects congressional votes or appropriations.

A more reasonable goal for the agency is to raise the level of debate on highly complex issues, framing and elucidating those factors that are important, stripping away those that are not, and focusing attention on those values and assumptions that lead to different policy preferences. Such a study would clarify the choices facing Congress and the consequences of those choices.

Occasionally, a study might aspire to do even more. Bringing diverse stakeholders together, allowing them to build a better understanding of the issues and each other's perspectives, and providing them with new analyses might reveal additional choices or permit them to reevaluate positions. Such a study could create a consensus where none had existed before.

A third possible outcome for a study—the "boring" study that, despite a competent effort, ultimately breaks no new ground—should also be recognized as a success. A scientific experiment that produces a null result has nevertheless generated useful information by ruling out a hypothesized approach or an otherwise plausible mechanism. In the same way, negative results from a competently executed study may show that an issue resists clarification, has been improperly framed, or is not yet ripe for public policy action. Institutions that cannot tolerate boring results are forced instead to sensationalize irrelevant ones, risking not only wasting resources but the more serious danger of misleading their clients.

Further Challenges

An advisory agency organized along the lines of this model faces a number of additional challenges, listed here along with suggestions on how best to address them.

Countering any tendency for the agency to serve primarily the needs of the majority party. The agency's bipartisan, bicameral oversight committee would be responsible for protecting it from slighting the interests of the minority party. Because the oversight committee is evenly divided between the two parties, no matter what the overall composition of Congress, the minority party will have stronger representation on the oversight committee than it has in the corresponding chamber. Therefore, the minority members of the oversight committee should be in a position to prevent majority party members from hijacking the agency to pursue a partisan agenda.

Responding to congressional demand for scientific and technological advice in a timely manner. The complexity of the large-scale, technology-intensive issues addressed by this agency, and the procedures necessary to ensure balance and accuracy, will place limits on how rapidly the agency will be able to produce major analyses. At the same time, if serious attention is given to Congress's need for timely response, many problems can be subdivided into more modest subsets, thus allowing the timely production of interim results. Furthermore, once a major assessment has been completed in some area of public policy, it should provide a base of expertise and a record of credibility that would allow the agency to deliver shorter term responses to related questions without going through its entire formal assessment process. The nature of these responses and the degree to which they could extend beyond the agency's previous work would be negotiated with its requesters and its overseers to ensure that the responses met the timing and other constraints of the congressional process while still satisfying the agency's requirements for accuracy and objectivity. At the same time, the agency would have to ensure that such short-term work did not systematically displace the long-term, in-depth, interdisciplinary analyses that would form the intellectual and programmatic core of the agency's efforts.

After all, the nature of the science and technology-intensive issues put before this agency will affect the time scale on which Congress needs answers. Few of these problems emerge instantaneously; instead, they tend to be chronic societal issues that return to the congressional agenda again and again. Such issues are rarely amenable to quick fixes. With appropriate foresight, the agency will have time to complete a major assessment on such a topic before Congress finally comes to closure. Although the ability of this agency to provide quick responses in areas in which it has recently worked will be important to meet congressional needs, the time needed for Congress to resolve chronic issues depends far more on the difficulty of reconciling society's diverse and competing interests than it does on waiting for scientific and technological analysis.

Avoiding the appearance or the reality that the agency's staff will advocate particular policy preferences, turning the agency into another lobbying organization or interest group. The agency must depend on its process and its review procedures to avoid bias. Although it would be inappropriate to select staff on the basis of their personal views on public policy topics, it would be desirable for agency staff to reflect a diversity of views and backgrounds. Moreover, agency staff should avoid taking public positions on policy questions that would appear to impugn their own impartiality.

Countering the belief that Congress should get science advice directly from the nation's best scientists. In Chapter 4, David H. Guston provides an excellent refutation of this position. Advice provided directly from scientists to policymakers is likely to be given privately and may therefore be politicized. Private advice would also not enjoy the peer review and public vetting that are essential for good governance as well as quality control. Most significantly, the best scientific or technological experts, no matter how strong their individual expertise, will not necessarily be able to address the broader, interdisciplinary aspects of complex science and technology-intensive policy issues.

Nor are they equipped to evaluate the policy implications of technical principles and uncertainties, particularly as viewed through a diversity of societal perspectives. Congress is rarely asked to make purely technical judgements. For typical issues, scientific and technological considerations interact in a complex way with each other and with a host of additional factors. In such cases, this agency will be able to elicit the active cooperation of the nation's best minds and put that expertise in an appropriate public policy context for congressional consideration.

Countering the belief that a new science advisory mechanism is unnecessary, given the wealth of information that is already available to policymakers on practically any topic. This issue has been extensively discussed in previous chapters. In summary, this agency is needed to help Congress sort through, evaluate, and synthesize the glut of information; provide an authoritative assessment of its strengths, weaknesses, and biases; and fill the remaining gaps.

Implementation

The model we have outlined bears a striking resemblance to the Office of Technology Assessment (OTA). Given Congress's 1995 decision to shut OTA down, it is reasonable to wonder whether something like it could realistically be recreated. Despite a significant number of bipartisan cosponsors, legislation introduced to this end in the House during the 107th Congress in 2002 (H.R. 2148) died in committee, and its 108th Congress counterpart (H.R. 125) does not appear to be faring much better.

There are undoubtedly short-term political considerations, but in the long run the answer will depend on whether OTA was defunded because it suffered basic design flaws, was the victim of unique circumstances that have since changed, or was incompatible with the congressional environment in fundamental ways that make it impossible for it, or any similar institution, to succeed.

Although OTA was not perfect, we believe that the basic structure was well conceived. With several easily implemented improvements, such as a greater emphasis on the production of more timely studies and greater attention to serving the needs of the minority, we believe that a dedicated congressional support agency along the basic lines of OTA continues to be the best mechanism for providing science and technology advice to Congress.

12

An Independent Analysis Group That Works Exclusively for Congress, Operated by a Nongovernmental Organization

Caroline S. Wagner and
William A. Stiles Jr.

This chapter explores a model for providing objective, independent, and measured science and technology (S&T) advice and support to Congress through a nongovernmental organization (NGO) working exclusively for Congress but maintaining a separate and independent identity. One way to envision this model would be to think about the use by Congress of a contracting mechanism similar to a federally funded research and development center (FFRDC) used by federal agencies. The selection, operation, and oversight of this independent analysis group would be conducted in a manner similar to the processes used for selecting outside organizations in the model examined in Chapter 9.

Organization and Structure

The independent analysis group proposed in this chapter would operate most effectively if it were "hosted" by a larger, independent NGO that would provide administrative and other overhead support. Within this organization, Congress would sponsor and fund on a multiyear basis an analysis group that is dedicated to providing assessment and analytical support for Congress on issues with major science and technology components. The function and structure of this analysis group would be similar to those of FFRDCs used in the executive branch. FFRDCs operate under a special contracting relationship with the government governed by the Federal Acquisition Regulations. To discharge its responsibilities, an FFRDC requires access, beyond that common to the normal contractual relationship, to government data, including sensitive information. The FFRDC must conduct its business in a manner befitting its special relationship with the government: to operate in the public interest with objectivity and independence, to be free from organizational conflicts of interest, and to have full disclosure of its affairs to the sponsoring agency.

This approach has certain administrative advantages because FFRDC-like contracts usually run for a longer time than other contracts, perhaps as much as five years at a time, and the FFRDCs can be flexible and open-ended about what kinds of projects to take on during that time. Usually, funding is provided for the overall class of activities, not for specific projects. Whereas they have the advantages of being government contractors, FFRDCs also have the advantages of being private (usually nonprofit) entities. This model has advantages over other possible models for providing S&T advice to Congress:

- In contrast to a congressionally funded and operated body, the analysis group would be independent and free of ties that might be seen as biasing analysis to please the funding source. They would also be able to interact easily with industry, labor, nonprofit organizations, and other outside groups without coming under a number of the constraints imposed directly on congressional members and staff. At the same time, the close ties under this contract provide the analysis group access to sensitive government information.
- In contrast to the congressional staff model, this model has the advantage of being free of the impression of political bias. Handled correctly, the analysis group could become a trusted nonpartisan, independent source of technical information and advice.
- In contrast to an "academies" model, the analysis group would not be advocating for funding for itself or for the academy and therefore would be free of any indication that they were pushing for more funding for one particular area of science over another or for science as a whole in lieu of other uses of federal money.
- Finally, with a contract arrangement, there is the potential for better cost–quality ratios over an approach that envisions dedicated congressional staff. As pointed out at the Science and Technology Advice for Congress workshop on June 14, 2001, FFRDCs can be more expensive than other approaches, but they also yield higher quality results in many cases.

There are some political advantages of this approach as well. One is the ability of Congress to distance itself easily from the findings of the analysis group—to either criticize or embrace them—more than it would from an in-house entity. At the same time, such an analysis group funded on a multiyear basis, especially a period asynchronous with election cycles, would provide an institutional memory on important issues, even as members of Congress or congressional staff turn over.

Such an analysis group would need a base of funding sufficient to keep the organization up and running with a core staff. Any additional work imposed by Congress would have to be accompanied by sufficient funding to allow this entity to contract out for any additional research and produce the studies required by the sponsoring committee (something required of the old Office of Technology Assessment [OTA] as well). This would help keep the research entity vibrant and increase its credibility across a wide range of audiences.

Mission

The principal mission of the independent analysis group would be to interact with Congress to (1) anticipate congressional needs for input on decisions requiring scientific know-how or technical information, (2) provide objective analysis to Congress on those issues where science and technology can clarify options or where the technical nature of the question requires expertise not available from congressional staff, and (3) develop a longer term vision for Congress to allow them to see emerging issues over the horizon.

The challenge for the analysis group would be to maintain its core relationships with congressional committees and offices while exploiting as fully as possible the niche within which it is free to conduct business. OTA walked this fine line daily. The mission would be carried forward principally by conducting an agreed-on agenda of research and analysis, commissioned by Congress, and seeking the widest possible audience for the resulting findings, as appropriate. In doing so, the entity would have a diversified portfolio of activities, balanced along several dimensions, that broaden the scope of policy analysis. These activities include the following:

- a mix of projects that are both short-term or quite immediate in topicality along with those that are longer term or are intended to provide a look over the horizon. Short-term projects allow this entity to contribute to ongoing debates in a manner that accommodates the time constraints of committees and the legislative schedule. Long-term projects are intended to provide greater service to the policy process and to position both congressional staff and members as well as the group of analysts to be of greatest assistance when these issues heat up.
- a mix of science and engineering combined with social and political analysis. This puts the science and technology aspects within the social and political context in which Congress will encounter the issues.
- a balance between direct requests for work and self-initiated inquiry that reflects emerging issues for which there is insufficient visibility or no identified client committee. OTA also walked this fine line daily. The independence of OTA became an issue at various times.

The analysis group must be seen as having analytical rigor in its own right. Thus, long-term, self-directed or -selected studies would need to be a part of the mission of this organization. This group would need to achieve credibility on its own, independent of its relationship to Congress, to be seen by the sponsor and the public as providing independent and objective analysis.

Selection and Operation of the Outside Analysis Group

The NGO acting as a host for the independent analysis group would have to be carefully selected according to criteria laid out in authorizing legislation or in report language accompanying the appropriations for funding. Before proceeding with any contracting on this topic, it would be necessary to hold a number of hearings to determine the need, scope, and definition of activity of the analysis group envisioned. From this inquiry would come minimum standards of performance and definitions of qualifications that could be used in a request for proposals. The contract for the analysis group would be arranged for a fixed multiyear term for a minimum level of activity and a minimal staffing level. Some review of congressional contracting authority would be needed to ensure that this process could be accomplished. If sufficient authority did not exist, it would have to be included in authorizing legislation.

The evaluation and selection of contract proposals and the ensuing operation of the analysis group would best be conducted through a dedicated bipartisan, bicameral body, using the joint congressional committee model. Such a congressional body could be established within the authorization bill or report language authorizing this function. If a joint entity were not established, alternatively a method could be used to obtain concurrent approval of the House Administration Committee and the Senate Committee on Rules and Administration.

Requests for work could be done in several ways. One approach would be to establish a research plan at the beginning of the fiscal year that would reflect conversations between the research and congressional staffs intended to identify important issue areas that warrant analytical study. This approach would allow the research staff to plan resource allocation and identify research analysts with backgrounds appropriate to important issues. For example, if energy issues are considered important to analyze in a given year—even if the specific questions are not worked out—identifying this as a priority area would enable resources to be dedicated to it. Identifying this area would also enable research staff to see what other research groups might be working on important areas, thereby avoiding duplication of effort.

Another approach would be to work on a request-by-request basis. However, this approach might have several drawbacks. First, it may mean that an important area is ignored simply because it is politically difficult. Second, staffing issues become tricky when planners do not know what kinds of questions will be asked. Third, it is hard to know how to strike a balance among competing studies if no prior planning is done; simply working by request could result in one or two studies laying claim to the entire budget.

Whichever approach is used, it would be best if requests for work of the analysis group would require the approval of the relevant committee chair or chairs, with the approval of the ranking minority members to ensure a bipartisan and neutral approach. A small group of qualified congressional staff could be assigned to ensure that requests for work met the legal mandate, reached a minimum standard of quality and scope, and were fashioned in a manner that would afford maximum benefit. This model worked well for OTA.

A reserve fund could be provided within the legislative branch appropriations bills that would be drawn on when requests were approved. The fund would be set to accommodate a base number of requests of various sizes, and the size of the fund would be revisited each appropriations cycle. Such a fund would provide the resources needed to move beyond the minimal staffing level to contract out for the work required and to publish or post the results of that research. In addition, some procedure would have to be established to adjudicate among committees making requests and to ensure that the requests approved did not exceed the resources provided. When requests did exceed resources, though, the additional work would be contingent on funding.

On completion of the requested or commissioned work, the staff of the congressional body, together with the requesting committee staff, would review the final publication and coordinate its release. Before release, reports would need to pass through a peer-review process. Depending on the NGO operating the research entity, a university for example, may be able to draw on internal expertise to provide peer review. Review can be conducted by a committee, along the National Academy of Sciences model, or by independent experts, similar to the method used by research journals. Peer reviewers might be nominated by staff from the bicameral, bipartisan committee to ensure that various opinions have been heard. Before publication, it may be useful to have the study results briefed to several important Congress members and possibly others to get feedback and identify problems with the delivery of the message. There would have to be some standards worked out to define the power of these staff and committee members to veto or significantly change the results of the work performed.

It may be useful to have a buffer built in, perhaps exercised by the bipartisan, bicameral body, to shield the analysis group from criticism in cases where it provides observations and information to Congress that prove to be antithetical to the majority view. This would help ensure independence and ensure that the results of the work performed would not be buried if they were contrary to the wishes of the requesting committee or the body overseeing the contracts.

Budget

The independent analysis group would need a budget that would support a core staff of perhaps 20 full-time equivalent positions while also allowing active use of outside expert consultants equaling perhaps another 40 full-time equivalent positions. Travel expenses, access to databases, and other costs associated with active research would also need to be included. However, the budget need not be as large as the former OTA budget to be effective, although the critical mass of the analysis group remains an important question. If broad expertise is not available within or through this organization, the organization will lose the confidence of its sponsor. An examination of the budgets of a range of similar research entities would be instructive.

Budgeting and contracting for the analysis group would need special consideration because adding the entity to Congress's budget would be a cause for some political debate. The best approach would be to have the budget line appear in the two relevant authorization committees, the House Administration Committee and the Senate Committee on Rules and Administration, and the appropriations would appear in the legislative branch appropriations bill. Because congressional outlays were a factor in the demise of the original OTA, it is probably best to address this issue head-on and not try to hide the budget as a "pass-through" in some other committee or agency budget. Asking another federal organization to contract for this function as a budget "pass-through" is quite unpopular among agencies asked to do this and might affect the perception of objectivity of the research entity.

Discussion

The challenge facing Congress is not one of insufficient information, even scientific and technological information. Rather it is one of analysis and synthesis of the information that flows to Congress. In the charged political atmosphere of congressional deliberations, "facts" become defined relatively: the true "facts" are the ones that support the position of the requesting member. The neutrality that is being sought in creating new mechanisms to provide science and technology advice to Congress is arguably one of the elements that caused the old OTA to lose supporters. This fine line needs to be trod carefully. More than one member voted against OTA because it raised fundamental questions about the feasibility of the Strategic Defense Initiative ("Star Wars") plan in a study stating that the technology and software were not ready for deployment.

Maintaining this balance between objectivity and relevance is a delicate one. "Speaking truth to power," as Wildavsky (1979) has pointed out, is always a risky business. Having a director of this group who has the stature to speak openly and perhaps even confidentially to Congress would be one step in aiding the process of providing advice. Similarly, like OTA, this group should not make specific policy recommendations, but rather describe a range of policy options. Primarily, the focus should be on infusing the legislative process with needed and relevant scientific or technical information when most other sources of such information are not free of bias.

Finally, the independent analysis group should be able to direct some of its research, and this work should be peer-reviewed and published. This work will help the group gain the confidence of the S&T community, and the analysis group can thereby become a trusted gateway for information transfer between Congress and the S&T community.

Reference

Wildavsky, Aaron B. 1979. *Speaking Truth to Power: The Art and Craft of Policy Analysis.* New York: Little Brown and Company.

Part IV

Moving toward Solution

13

Where Do We Go from Here?

M. Granger Morgan and Jon M. Peha

America's founding fathers created a Congress of citizen-legislators that uses a decisionmaking model derived from the adversarial traditions of the law. Under this model, members listen to the arguments and pleadings of interested parties on all sides of an issue, and from the resulting synthesis and balancing of interests they collectively discern and act on the national interest.

In Chapter 1 we argued that, whereas this model works well for addressing many problems, an important and growing class of problems involves complex issues of science and technology, in which this traditional model needs to be augmented with systematic analysis and synthesis if Congress and its members are to make wise, well-informed decisions. For example, how can critical infrastructures such as electric power systems and computer networks best be made robust and reliable while continuing to operate in a competitive environment in which most investment decisions are driven by short-term market considerations? What are the risks, costs, and benefits of systems to separate carbon from fossil fuels and sequester it deep underground where it cannot reach the atmosphere? And what role should government play in the development of these technologies? How can the potential risks of new therapeutic drugs and medical devices best be balanced against the lifesaving potential of rapid introduction? How can the explosive growth in wireless personal electronic devices be managed so that it does not adversely affect the safety of commercial airliners? The list goes on.

The current sources of independent, balanced analytical support that Congress can routinely command do not span the full range of response times. There is a gap for studies that require more time than a few weeks but less

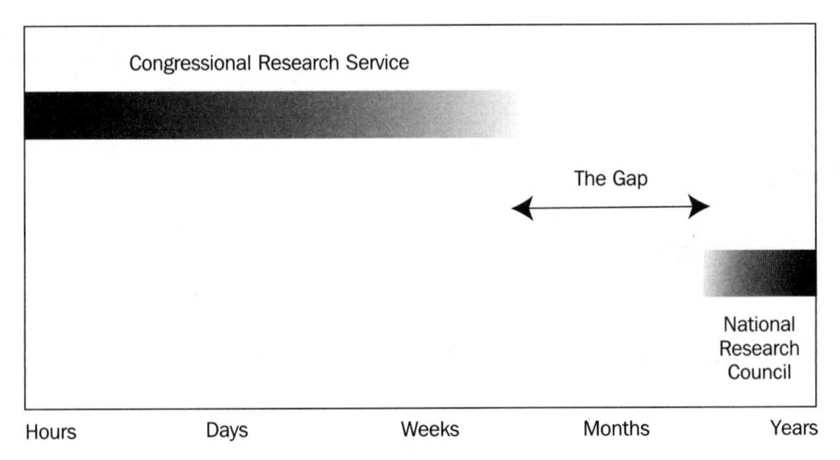

Figure 13-1. Sources and Response Times of Analysis Available to Congress

Note: The current sources of independent impartial analytical support that Congress can routinely command do not span the full range of response times. There is a gap for studies that require more time than a few weeks but less time than 18 months.

time than 18 months. The Congressional Research Service does an excellent job of providing short-term assistance. The National Research Council (NRC) is often used for longer term studies that, except in special circumstances, usually require a year or more. In between, there is a gap: Congress has no reliable, balanced, impartial sources of analysis and synthesis on policy problems involving issues of science and technology to which it can routinely turn for help on these intermediate time scales (Figure 13-1).

The sources of analysis currently available to Congress vary widely, both in terms of their balance, completeness, and impartiality and in terms of their sensitivity and responsiveness to congressional needs. NRC reports are generally well balanced, although, because they are performed by experts, they may not always obtain input from or consider the views of as wide a set of stakeholders as Congress must consider. Often too, rather than lay out a range of policy options with a discussion of their strengths and limitations, NRC reports make specific policy recommendations. The result is that it can be difficult for Congress to use such reports in making judgements and informed political and technical trade-offs.

Analysis performed by universities and think tanks can be extremely valuable, but it can be uneven in coverage, balance, and responsiveness to congressional needs, and it is often not available on topics or at times that meet congressional needs. Furthermore, it can be difficult for members and staff to assess how balanced, complete, and impartial such work is. Peer-reviewed publication of the findings can help, but such review does not assess all the factors that Congress must consider. Analytical input from interest groups can be highly responsive to congressional needs but is often incomplete or based on biased assumptions, and it tells only part of the relevant story.

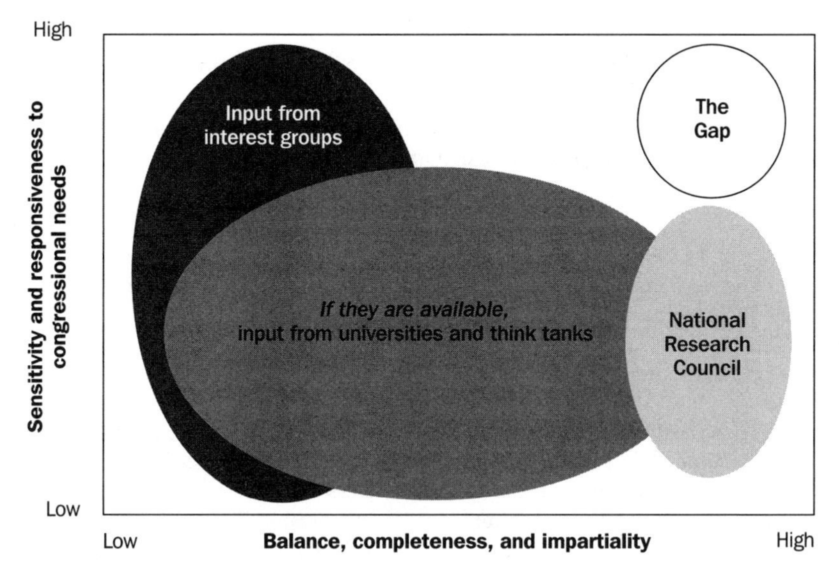

Figure 13-2. Analytical Capability Available to Congress

Note: The sources of analysis currently available to Congress vary widely both in terms of their balance, completeness, and impartiality and in terms of their sensitivity and responsiveness to the needs of Congress. As with study time (Figure 13-1), there is a gap for analysis that scores well on both dimensions.

Just as with study duration, there is a gap in analytical capability that is simultaneously balanced, complete, and impartial while also sensitive and responsive to the needs of Congress. This second gap is illustrated graphically in Figure 13-2.

Congress is a representative body that responds to its constituents. Most members do not have technical or policy–analytic backgrounds. Several members who have such backgrounds have argued to us that if Congress is going to create institutional arrangements to fill the gaps shown in Figures 13-1 and 13-2, outside constituents will have to mobilize to persuade Congress that this capability is sorely needed and that its addition will lead to more informed decisions that better serve the public interest. Absent such external encouragement, the majority of members and congressional staff are unlikely to take action to make such institutional changes.

Budget limitations and other considerations may lead Congress to choose to develop this analytical capability incrementally. If this happens, there are likely to be many opportunities to make either good or bad decisions, so it is important for the technical community to remain attentive and engaged.

From the discussion of the preceding chapters, we conclude that several different institutional arrangements could serve congressional needs. In Table 13-1, we have summarized what we and our co-authors see as the strengths and weaknesses of the various models we have considered. All of these models could provide Congress with analytical support. Some combination of the fea-

Table 13-1. Comparison of Some of the Advantages and Disadvantages of the Alternative Institutional Models for Providing Science and Technology Analytical Advice to the U.S. Congress

***Institutional model:* An expanded analytical capability in the Congressional Research Service, the General Accounting Office, or the Congressional Budget Office**

Disadvantages	*Advantages*
• CRS, GAO, and CBO have other missions and have institutional cultures and incentive structures that may not be compatible with developing a sustained ability to perform interdisciplinary policy analysis. • The bulk of the costs would appear on the legislative branch budget. • This might result in squeezing the budgets of these organizations that are available to support their traditional missions.	• This is an incremental change to an existing and familiar mechanism. • The capability is entirely within the legislative branch, and thus completely under its control and dedicated to its needs. • These organizations have extensive knowledge about the needs of Congress. • Properly organized, this system could maintain tight bipartisan, bicameral control over the process and systematic review of the work products. • Properly organized, it could provide a range of other technical advice and support to Congress. • At least in the case of GAO, the capability to begin to make the necessary institutional and cultural changes has been demonstrated.

***Institutional model:* Expanded use of the National Academies complex**

Disadvantages	*Advantages*
• Expanded use does not maintain tight bipartisan, bicameral control over the process or the systematic review of the work products. • This option is more likely to produce products that "tell the Congress what to do" rather than simply frame the problem and lay out options for consideration. • It poses some political risks to the NAS–NRC complex that might undermine their ability to perform their current roles. • It provides a bit more distance between the analytical organization and Congress, making it easier to trim or eliminate the budget.	• This is an incremental change to an existing and familiar mechanism. • This organization has very high status and prestige. • The complex provides a bit more distance between the analytical organization and Congress, perhaps making it less vulnerable to being attacked for unwanted messages. • It might be possible to move some of the costs off the legislative branch budget.

continued on next page

Table 13-1. Comparison of Some of the Advantages and Disadvantages of the Alternative Institutional Models for Providing Science and Technology Analytical Advice to the U.S. Congress *(continued)*

Disadvantages	*Advantages*
***Institutional model:* Expanding the role of the science and engineering congressional fellowship program**	
• Because fellows are located on the personal staffs of members or on majority or minority staffs of committees, framing of analysis requests may be influenced by partisan considerations. • Support for fellows is already limited, and this might place further demands on an already tight resource base and on scarce time of fellows.	• This is an incremental change to an existing and familiar mechanism. • This model engages the services of staff who have substantial technical expertise in the commissioning and oversight of analysis.
***Institutional model:* A lean, distributed organization to serve Congress**	
• This organization involves acquisition of outside resources and thus is subject to the usual political incentives to spread the wealth around. • It involves analysis performed by staff not directly in control of Congress and thus perhaps less sensitized to unique congressional needs and sensitivities. • It provides a bit more distance between the analytical organization and Congress, making it easier to trim or eliminate the budget.	• This organization maintains tight bipartisan, bicameral control over the process and systematic review of the work products. • It offers a way to minimize the size of the organization within the legislative branch while maintaining great analytical depth and capability. • It provides a bit more distance between the analytical organization and Congress, perhaps making it less vulnerable to being attacked for unwanted messages. • It might be possible to move some of the costs off the legislative branch budget.
***Institutional model:* A dedicated organization in Congress**	
• This organization is similar to the old OTA, and thus perhaps politically objectionable to some members. • The bulk of the costs would appear on the legislative branch budget. • This organization must balance the need to employ full-time experts in a wide range of disciplines with the need to limit the budget. • As a free-standing unit, this organization is more vulnerable to shifts in political climate.	• This organization is entirely within the legislative branch, and thus completely under its control and dedicated to its needs. • It maintains tight bipartisan, bicameral control over the process and systematic review of the work products. • Properly organized, it could provide a range of other technical advice and support to Congress.

continued on next page

Table 13-1. Comparison of Some of the Advantages and Disadvantages of the Alternative Institutional Models for Providing Science and Technology Analytical Advice to the U.S. Congress *(continued)*

***Institutional model:* A dedicated organization outside Congress**	
Disadvantages	*Advantages*
• With this organization, it may be harder to maintain tight bipartisan, bicameral control over the process or the systematic review of the work products. • This process involves acquisition of outside resources and thus is subject to the usual political incentives to spread the wealth around. • It provides a bit more distance between the analytical organization and Congress, making it easier to trim or eliminate the budget. • It involves analysis performed by staff not directly in control of Congress and thus perhaps less sensitized to unique congressional needs and sensitivities.	• This organization provides a bit more distance between the analytical organization and Congress, perhaps making it less vulnerable to being attacked for unwanted messages. • It might be possible to move some of the costs off the legislative branch budget.

tures of two or three might be better than any one of them as they have been described in this book. However, given the complexities and vagaries of congressional decisionmaking, one should not aim too single-mindedly for any specific optimal arrangement but rather should encourage a range of activities and undertakings that together advance the goal of providing more and better analytical input for Congress on issues involving science and technology. If political or budget considerations preclude adoption of the best solution, any of a number of alternatives would be better than allowing the current gaps to persist.

We believe that the need for improved analysis would be best served by an analysis organization that works exclusively for Congress and is located inside the legislative branch under the control of a bipartisan, bicameral committee. Such a group could either be free-standing (the models advanced in Chapters 10 and 11) or could be housed as part of some existing legislative branch organization such as the General Accounting Office (GAO) (the model discussed in Chapter 7 and Appendix 3).

Because the limited size of such an analysis group will preclude it from having the full range of specialized expertise it will need, there is a big advantage to its being able to draw on the special expertise of outside organizations such as universities and nonprofit think tanks. This implies a combination of features drawn from the models outlined in Chapters 8, 10, and 12. The organization could operate effectively with various mixtures of in-house and external expertise, but at least some in-house substantive and analytical expertise is highly desirable. Arrangements in which most or all of the expertise resides outside Congress itself are less likely to build the institutional commitment or provide the informal support services that a sustained analytical presence will need if it is to adequately serve the ever-changing environment of Congress.

The cost of such a proposed analytical capability would be modest, on the order of $10–30 million per year. This would amount to between 0.35% and 1% of the total budget of the legislative branch, and 0.0005% and 0.0015% of the total federal budget—a minuscule expense when compared with the costs associated with the decisions that its analysis would be used to inform and support.

Although the existing legal framework for the former Office of Technology Assessment—including control by a bipartisan, bicameral committee—could be used to establish such a capability, if that strategy were pursued and it gained general political support, some improvements should be made with respect to the design and operation of the new unit.[1] Among these are strategies to:

- offer a more diversified set of study options capable of producing many studies in a more timely manner;
- ensure that the needs of the minority party are well served;
- use modern interactive capabilities (such as Web-based dialog, workshops, and short-term studies); and
- more explicitly supply technical advice and support to other congressional support organizations and to individual members.

Given the political baggage that is associated with the name of the Office of Technology Assessment (OTA), and the fact that the new organization should adopt new strategies, we believe that it would be wise to choose a new name, even if the original Technology Assessment Act were used as the basic implementing framework. We suggest either the "Congressional Office of Technology Analysis" or the "Congressional Office of Science and Technology Analysis."

The former OTA's official function was to do analysis for committees, but it actually played a more complex role. Its reports were widely used by interest groups, academics, and the general public.[2] By laying out problems and a range of possible solutions, the reports helped to inform a wide range of national debate. Many participants in the workshop held on this topic in Washington, D.C., in June 2001, as well as others inside and outside Congress with whom we have discussed these issues, have argued that sometimes the most important effects of OTA studies did not come from direct inputs provided to committees and members. They came, instead, through informed feedback from constituents who used OTA reports to frame and support their arguments on all sides of an issue. Sometimes OTA performed studies of issues well before they got onto the active political agenda. When this happened, their reports helped to provide structure to the political discourse when the issue later became "hot."

There is much to favor the creation of a free-standing analysis group within Congress, but the successful completion of the technology assessment *Using*

Biometrics for Border Security by the GAO (2002), demonstrates that, with some further refinement, this function could also be effectively performed within that existing organization. Whereas the initial GAO effort did not do as well as it might have in terms of adopting a broad policy–analytic perspective, this shortcoming could be rectified if GAO takes steps to strengthen its policy–analytic staff capabilities, makes appropriate changes in internal guidance and administrative arrangements, and, because small programs cannot develop all the necessary expertise in-house, makes effective use of outside expertise and contractors (Fri et al. 2002). Appendix 3 provides additional details on these points.

We believe that it is desirable to establish a new analysis group within Congress, but we also believe that steps should be taken to strengthen some of the existing institutional mechanisms that serve congressional needs. Desirable changes include the following:

- Expanding and strengthening the role of the National Academies complex (the National Academy of Science, the National Academy of Engineering, the Institute of Medicine, and the National Research Council) and encouraging the creation of a class of studies that are fast and that structure a problem and lay out policy options for Congress but do not tell Congress what it should do.
- Strengthening the technical and policy–analytic capabilities of the existing congressional support agencies. This is needed even if one of these agencies, such as GAO, is not ultimately used to house a new analysis group. The existence of expanded capabilities in these organizations would provide valuable institutional redundancy.
- Providing members of legislative staff in both committee and personal offices with information on how to secure more systematic access to external resources, including groups such as universities or think tanks as well as any new congressional analysis organization. Seminars might be held during recess to inform staff members about available resources and how to use them. These would be similar to the sessions offered by the Congressional Research Service on other topics. Because American Association for the Advancement of Science (AAAS) fellows are particularly likely to make effective use of these resources, similar sessions could be added to the AAAS orientation program.

Americans are typically reluctant to look abroad for institutional inspiration. Furthermore, institutions that work well in one national and cultural setting can rarely be transplanted without change to another. However, as Chapter 5 makes clear, the original OTA provided the inspiration for the creation of science and technology analysis groups that now serve 15 European parliaments. None of these institutions looks exactly like the old OTA, and none could be appropriately retransplanted back to the U.S. Congress. However, the existence of these bodies provides a strong international endorsement of the insight that in today's complex world, legislative bodies need analytical assistance when they deal with complex policy issues involving science and technology.

Drawing on the lessons of the past and the experience of others, it is time for the U.S. Congress to once again establish a set of institutional arrangements by which it can obtain balanced, independent analytical advice and guidance on complex policy issues that involve science and technology.

Note

[1] Making changes to the Technology Assessment Act of 1972, as opposed to simply appropriating new funds under the existing legal framework, could add complications. In a supportive political environment, these could be readily overcome. In a hostile political environment, they could be used to prevent the effort from proceeding.

[2]Copies of the OTA's reports are available on CD from the U.S. Government Printing Office (for details, see www.access.gpo.gov/ota/ [accessed May 3, 2003]) and are available on line at www.wws.princeton.edu/~ota/ (accessed May 3, 2003).

References

Fri, Robert, M. Granger Morgan, and William A. (Skip) Stiles Jr. 2002. *An External Evaluation of GAO's Assessment of Technologies for Control.* Washington, DC: General Accounting Office.

General Accounting Office (GAO). 2002. *Technology Assessment: Using Biometrics for Border Security.* GAO–03–174. Washington, DC: General Accounting Office.

Appendix 1

The Technology Assessment Act of 1972

Public Law 92–484
92d Congress, H.R. 10243
October 13, 1972

An Act

To establish an Office of Technology Assessment for the Congress as an aid in the identification and consideration of existing and probable impacts of technological application; to amend the National Science Foundation Act of 1950; and for other purposes. Be it enacted by the Senate and House of Representatives of the United States of America in Congress assembled, That this Act may be cited as the Technology Assessment Act of 1972.

Findings and Declaration of Purpose

SEC. 2. The Congress hereby finds and declares that:

(a) As technology continues to change and expand rapidly, its applications are—
   1. large and growing in scale; and
   2. increasingly extensive, pervasive, and critical in their impact, beneficial and adverse, on the natural and social environment.

(b) Therefore, it is essential that, to the fullest extent possible, the consequences of technological applications be anticipated, understood, and considered in determination of public policy on existing and emerging national problems.

(c) The Congress further finds that:
1. the Federal agencies presently responsible directly to the Congress are not designed to provide the Legislative Branch with adequate and timely information, independently developed, relating to the potential impact of technological applications, and
2. the present mechanisms of the Congress do not and are not designed to provide the Legislative Branch with such information.

(d) Accordingly, it is necessary for the Congress to—
1. equip itself with new and effective means for securing competent, unbiased information concerning the physical, biological, economic, social, and political effects of such applications; and
2. utilize this information, whenever appropriate, as one factor in the legislative assessment of matters pending before the Congress, particularly in those instances where the Federal Government may be called upon to consider support for, or management or regulation of, technological applications.

Establishment of the Office of Technology Assessment

SEC. 3. (a) In accordance with the findings and declaration of purpose in section 2, there is hereby created the Office of Technology Assessment (hereinafter referred to as the Office) which shall be within and responsible to the legislative branch of the Government.

(b) The Office shall consist of a Technology Assessment Board (hereinafter referred to as the Board) which shall formulate and promulgate the policies of the Office, and a Director who shall carry out such policies and administer the operations of the Office.

(c) The basic function of the Office shall be to provide early indications of the probable beneficial and adverse impacts of the applications of technology and to develop other coordinate information which may assist the Congress. In carrying out such function, the Office shall:
1. identify existing or probable impacts of technology or technological programs;
2. where possible, ascertain cause-and-effect relationships;
3. identify alternative technological methods of implementing specific programs;
4. identify alternative programs for achieving requisite goals;
5. make estimates and comparisons of the impacts of alternative methods and programs;
6. present findings of completed analyses to the appropriate legislative authorities;
7. identify areas where additional research or data collection is required to provide adequate support for the assessments and estimates described in paragraph (1) through (5) of this subsection; and

8. undertake such additional associated activities as the appropriate authorities specified under subsection (d) may direct.

(d) Assessment activities undertaken by the Office may be initiated upon the request of:
1. the chairman of any standing, special, or select committee of either House of the Congress, or of any joint committee of the Congress, acting for himself or at the request of the ranking minority member or a majority of the committee members;
2. the Board; or
3. the Director, in consultation with the Board.

(e) Assessments made by the Office, including information, surveys, studies, reports, and findings related thereto, shall be made available to the initiating committee or other appropriate committees of the Congress. In addition, any such information, surveys, studies, reports, and findings produced by the Office may be made available to the public except where
1. to do so would violate security statutes; or
2. the Board considers it necessary or advisable to withhold such information in accordance with one or more of the numbered paragraphs in section 552(b) of title 5, United States Code.

Technology Assessment Board

SEC. 4. (a) The Board shall consist of thirteen members as follows:
1. six Members of the Senate, appointed by the President pro tempore of the Senate, three from the majority party and three from the minority party;
2. six Members of the House of Representatives appointed by the Speaker of the House of Representatives, three from the majority party and three from the minority party; and
3. the Director, who shall not be a voting member.

(b) Vacancies in the membership of the Board shall not affect the power of the remaining members to execute the functions of the Board and shall be filled in the same manner as in the case of the original appointment.

(c) The Board shall select a chairman and a vice chairman from among its members at the beginning of each Congress. The vice chairman shall act in the place and stead of the chairman in the absence of the chairman. The chairmanship and the vice chairmanship shall alternate between the Senate and the House of Representatives with each Congress. The chairman during each even-numbered Congress shall be selected by the Members of the House of Representatives on the Board from among their number. The vice chairman during each Congress shall be chosen in the same manner from that House of Congress other than the House of Congress of which the chairman is a Member.

(d) The Board is authorized to sit and act at such places and times during the sessions, recesses, and adjourned periods of Congress, and upon a vote of a majority of its members, to require by subpoena or otherwise the attendance of such witnesses and the production of such books, papers, and documents, to administer such oaths and affirmations, to take such testimony, to procure such printing and binding, and to make such expenditures, as it deems advisable. The Board may make such rules respecting its organization and procedures as it deems necessary, except that no recommendation shall be reported from the Board unless a majority of the Board assent. Subpoenas may be issued over the signature of the chairman of the Board or of any voting member designated by him or by the Board, and may be served by such person or persons as may be designated by such chairman or member. The chairman of the Board or any voting member thereof may administer oaths or affirmations to witnesses.

Director and Deputy Director

SEC. 5. (a) The Director of the Office of Technology Assessment shall be appointed by the Board and shall serve for a term of six years unless sooner removed by the Board. He shall receive basic pay at the rate provided for level III of the Executive Schedule under section 5314 of title 5, United States Code.

(b) In addition to the powers and duties vested in him by this Act, the Director shall exercise such powers and duties as may be delegated to him by the Board.

(c) The Director may appoint with the approval of the Board, a Deputy Director who shall perform such functions as the Director may prescribe and who shall be Acting Director during the absence or incapacity of the Director or in the event of a vacancy in the office of Director. The Deputy Director shall receive basic pay at the rate provided for level IV of the Executive Schedule under section 5315 of title 5, United States Code.

(d) Neither the Director nor the Deputy Director shall engage in any other business, vocation, or employment than that of serving as such Director or Deputy Director, as the case may be; nor shall the Director or Deputy Director, except with the approval of the Board, hold any office in, or act in any capacity for, any organization, agency, or institution with which the Office makes any contract or other arrangement under this Act.

Authority of the Office

SEC. 6. (a) The Office shall have the authority, within the limits of available appropriations, to do all things necessary to carry out the provisions of this Act, including, but without being limited to, the authority to

1. make full use of competent personnel and organizations outside the Office, public or private, and form special ad hoc task forces or make other arrangements when appropriate;
2. enter into contracts or other arrangements as may be necessary for the conduct of the work of the Office with any agency or instrumentality of

the United States, with any State, territory, or possession or any political subdivision thereof, or with any person, firm, association, corporation, or educational institution, with or without reimbursement, without performance or other bonds, and without regard to section 3709 of the Revised Statutes (41 U.S.C. 5);

3. make advance, progress, and other payments which relate to technology assessment without regard to the provisions of section 3648 of the Revised Statutes (31 U.S.C. 529);
4. accept and utilize the services of voluntary and uncompensated personnel necessary for the conduct of the work of the Office and provide transportation and subsistence as authorized by section 5703 of title 5, United States Code, for persons serving without compensation;
5. acquire by purchase, lease, loan, or gift, and hold and dispose of by sale, lease, or loan, real and personal property of all kinds necessary for or resulting from the exercise of authority granted by this Act; and
6. prescribe such rules and regulations as it deems necessary governing the operation and organization of the Office.

(b) Contractors and other parties entering into contracts and other arrangements under this section which involve costs to the Government shall maintain such books and related records as will facilitate an effective audit in such detail and in such manner as shall be prescribed by the Office, and such books and records (and related documents and papers) shall be available to the Office and the Comptroller General of the United States, or any of their duly authorized representatives, for the purpose of audit and examination.

(c) The Office, in carrying out the provisions of this Act, shall not, itself, operate any laboratories, pilot plants, or test facilities.

(d) The Office is authorized to secure directly from any executive department or agency information, suggestions, estimates, statistics, and technical assistance for the purpose of carrying out its functions under this Act. Each such executive department or agency shall furnish the information, suggestions, estimates, statistics, and technical assistance directly to the Office upon its request.

(e) On request of the Office, the head of any executive department or agency may detail, with or without reimbursement, any of its personnel to assist the Office in carrying out its functions under this Act.

(f) The Director shall, in accordance with such policies as the Board shall prescribe, appoint and fix the compensation of such personnel as may be necessary to carry out the provisions of this Act.

Establishment of the Technology Assessment Advisory Council

SEC. 7. (a) The Office shall establish a Technology Assessment Advisory Council (hereinafter referred to as the Council). The Council shall be composed of the following twelve members:

1. ten members from the public, to be appointed by the Board, who shall be persons eminent in one or more fields of the physical, biological, or social sciences or engineering or experienced in the administration of technological activities, or who may be judged qualified on the basis of contributions made to educational or public activities;
2. the Comptroller General; and
3. the Director of the Congressional Research Service of the Library of Congress.

(b) The Council, upon request by the Board, shall—

1. review and make recommendations to the Board on activities undertaken by the Office or on the initiation thereof in accordance with section 3(d);
2. review and make recommendations to the Board on the findings of any assessment made by or for the Office; and
3. undertake such additional related tasks as the Board may direct.

(c) The Council, by majority vote, shall elect from its members appointed under subsection (a)(1) of this section a Chairman and a Vice Chairman, who shall serve for such time and under such conditions as the Council may prescribe. In the absence of the Chairman, or in the event of his incapacity, the Vice Chairman shall act as Chairman.

(d) The term of office of each member of the Council appointed under subsection (a)(1) shall be four years except that any such member appointed to fill a vacancy occurring prior to the expiration of the term for which his predecessor was appointed shall be appointed for the remainder of such term. No person shall be appointed a member of the Council under subsection (a)(1) more than twice. Terms of the members appointed under subsection (a)(1) shall be staggered so as to establish a rotating membership according to such method as the Board may devise.

(e)

1. The members of the Council other than those appointed under subsection (a)(1) shall receive no pay for their services as members of the Council, but shall be allowed necessary travel expenses (or, in the alternative, mileage for use of privately owned vehicles and per diem in lieu of subsistence at not to exceed the rate prescribed in sections 5702 and 5704 of title 5, United States Code), and other necessary expenses incurred by them in the performance of duties vested in the Council, without regard to the provisions of subchapter 1 of chapter 57 and section 5731 of title 5, United States Code, and regulations promulgated thereunder.
2. The members of the Council appointed under subsection (a)(1) shall receive compensation for each day engaged in the actual performance of duties vested in the Council at rates of pay not in excess of the daily equivalent of the highest rate of basic pay set forth in the General Schedule of section 5332(a) of title 5, United States Code, and in addition shall be reimbursed for travel, subsistence, and other necessary expenses in the manner provided for other members of the Council under paragraph (1) of this subsection.

Utilization of the Library of Congress

SEC. 8. (a) To carry out the objectives of this Act, the Librarian of Congress is authorized to make available to the Office such services and assistance of the Congressional Research Service as may be appropriate and feasible.

(b) Such services and assistance made available to the Office shall include, but not be limited to, all of the services and assistance which the Congressional Research Service is otherwise authorized to provide to the Congress.

(c) Nothing in this section shall alter or modify any services or responsibilities, other than those performed for the Office, which the Congressional Research Service under law performs for or on behalf of the Congress. The Librarian is, however, authorized to establish within the Congressional Research Service such additional divisions, groups, or other organizational entities as may be necessary to carry out the purpose of this Act.

(d) Services and assistance made available to the Office by the Congressional Research Service in accordance with this section may be provided with or without reimbursement from funds of the Office, as agreed upon by the Board and the Librarian of Congress.

Utilization of the General Accounting Office

SEC. 9. (a) Financial and administrative services (including those related to budgeting, accounting, financial reporting, personnel, and procurement) and such other services as may be appropriate shall be provided the Office by the General Accounting Office.

(b) Such services and assistance to the Office shall include, but not be limited to, all of the services and assistance which the General Accounting Office is otherwise authorized to provide to the Congress.

(c) Nothing in this section shall alter or modify any services or responsibilities, other than those performed for the Office, which the General Accounting Office under law performs for or on behalf of the Congress.

(d) Services and assistance made available to the Office by the General Accounting Office in accordance with this section may be provided with or without reimbursement from funds of the Office, as agreed upon by the Board and the Comptroller General.

Coordination with the National Science Foundation

SEC. 10. (a) The Office shall maintain a continuing liaison with the National Science Foundation with respect to—

1. grants and contracts formulated or activated by the Foundation which are for purposes of technology assessment; and
2. the promotion of coordination in areas of technology assessment, and the avoidance of unnecessary duplication or overlapping of research

activities in the development of technology assessment techniques and programs.

(b) Section 3(b) of the National Science Foundation Act of 1950, as amended (42 U.S.C. 1862(b)), is amended to read as follows:

"(b) The Foundation is authorized to initiate and support specific scientific activities in connection with matters relating to international cooperation, national security, and the effects of scientific applications upon society by making contracts or other arrangements (including grants, loans, and other forms of assistance) for the conduct of such activities. When initiated or supported pursuant to requests made by any other Federal department or agency, including the Office of Technology Assessment, such activities shall be financed whenever feasible from funds transferred to the Foundation by the requesting official as provided in section 14(g), and any such activities shall be unclassified and shall be identified by the Foundation as being undertaken at the request of the appropriate official."

Annual Report

SEC. 11. The Office shall submit to the Congress an annual report which shall include, but not be limited to, an evaluation of technology assessment techniques and identification, insofar as may be feasible, of technological areas and programs requiring future analysis. Such report shall be submitted not later than March 15 of each year.

Appropriations

SEC. 12. (a) To enable the Office to carry out its powers and duties, there is hereby authorized to be appropriated to the Office, out of any money in the Treasury not otherwise appropriated, not to exceed $5,000,000 in the aggregate for the two fiscal years ending June 30, 1973, and June 30, 1974, and thereafter such sums as may be necessary.

(b) Appropriations made pursuant to the authority provided in subsection (a) shall remain available for obligation, for expenditure, or for obligation and expenditure for such period or periods as may be specified in the Act making such appropriations.

Approved October 13, 1972.

Appendix 2

Details on the National Academies Complex

Background on the Complex

The National Academy of Sciences (NAS) is a private, nonprofit, self-perpetuating society of distinguished scholars engaged in scientific and engineering research, dedicated to the promotion of science and technology and to their uses for the general welfare. On the authority of the charter granted to it by Congress in 1863, the academy has a mandate that requires it to advise the federal government on scientific and technical matters. Bruce Alberts is currently serving his second term as president of the National Academy of Sciences and chair of the National Research Council (NRC).[1]

The National Academy of Engineering (NAE) was established in 1964, under the charter of the National Academy of Sciences, as a parallel organization of outstanding engineers. It is autonomous in its administration and in the selection of its members, sharing with the National Academy of Sciences the responsibility for advising the federal government. William A. Wulf began his second term as president of the National Academy of Engineering and vice chair of the NRC in April 2001.

The Institute of Medicine (IOM) was established in 1970 by the NAS to recognize distinguished professional achievement in medicine and health and to examine important issues related to human health. The institute acts under the responsibility given to the NAS by its congressional charter to be an adviser to the federal government and, upon its own initiative, to identify issues of medical care, research, and education. Harvey V. Fineberg began his first term as president of the IOM in July 2002.

For most of the past century, the principal focus on provision of scientific and engineering advice to government on the part of the National Academies complex has been through the NRC.

The NRC was organized by the National Academy of Sciences in 1916, and its purpose is to associate the broad community of science and technology with the NAS's purposes of furthering knowledge and advising the federal government. The NRC's study process is one in which a committee of scientific experts, with a diverse range of expertise and perspectives, is convened to address a particular set of questions.

Functioning in accordance with general policies determined by the NAS and the NAE, the NRC has become the principal operating agency of both the NAS and the NAE in providing services to the government, the public, and the scientific and engineering communities. Whereas the IOM is technically not part of the NRC, IOM studies also operate under the procedures of the NRC. The council is overseen by a governing board with members from the NAS, the NAE, and the IOM. Under current arrangements, NAS President Alberts has final approval authority for the membership of all NRC and IOM study committees. At any one time, almost 7,000 experts from science, engineering, and many other disciplines, including many from the NAS, the NAE, and the IOM, serve pro bono on NRC committees or boards or otherwise participate in NRC activities. The NRC is a mechanism for tapping broadly into the expertise of the scientific, engineering, and medical community outside of government; in fact 80% of the members of NRC committees are not members of the NAS, the NAE, or the IOM. Approximately 250 reports are produced each year, about half of which are consensus studies by expert committees.

NRC Congressional Mandates during the 102nd through 106th Congresses

102nd Congress: 40 Mandates

PL102–4 Agent Orange Act of 1991

- Study of scientific evidence linking certain diseases to exposure to dioxin and chemical compounds in other herbicides

PL102–104 Department of Energy and Water Appropriations

- Study of health risks associated with exposure to electromagnetic fields

PL102–136 Decennial Census Improvements Act of 1991

- Study for improving the decennial census

PL102–140 Departments of Commerce, Justice, State Appropriations Act FY1992

- Study of salmon in the Columbia River and recommendations to improve long-term sustainability of salmon (see also PL102–395)

PL102–154 Department of Interior Appropriations Act FY1992
- Study to assess forests in the Pacific Northwest and the relationship to national supply and demand for forest products

PL102–170 Departments of Labor, Health and Human Services, and Education Appropriations Act FY1992
- Study of Alcohol, Drug Abuse, and Mental Health Administration Reorganization Act (ADAMHA) AIDS program
- Study of women's health issues

PL102–172 Department of Defense Appropriations Act FY1992
- Study on United States–Soviet nuclear policy
- Study on safety, command, and control issues

PL102–190 National Defense Authorization Act FY1992 and FY1993
- Study to assist the U.S. Navy in conducting mines countermeasure technology
- Study to analyze the strengths and weaknesses in Japanese science and technology, including a framework for scientific technology cooperation with Japan

PL102–237 Food, Agriculture, Conservation, and Trade Act Amendments of 1991
- Study to develop criteria for and evaluate current and future inspection exemptions for meat and poultry food products

PL102–240 Intermodal Surface Transportation Efficiency Act of 1991
- Study of a cooperative transit research program
- Study of current and potential impediments to international standards in intermodal transportation
- Study of data collection procedures and capabilities of the Department of Transportation

PL102–245 American Technology Preeminence Act of 1991
- Study of international product testing, certification, and quality control issues related to international trade

PL102–321 Alcohol, Drug Abuse, and Mental Health Administration Reorganization Act (ADAMHA)
- Study of antiaddiction medications
- Study of fetal alcohol syndrome
- Study to evaluate programs in the United States that provide sterile hypodermic needles and bleach to reduce the risk of contracting AIDS
- Study to assess statutory allotment formulae for funds made available to states and territories for substance abuse and mental health programs

PL102–325 Higher Education Amendments of 1992
- Study of civilian aviation training programs needed by the commercial aviation industry in the year 2000
- Study of the research program of teacher assessments carried out by the National Board for Professional Teaching Standards
- Study on graduate education information
- Study of the annual report of the National Board for Professional Teaching Standards

PL102–368 Supplemental Appropriations Act FY1993
- Study of the investigative strategy of the National Institute for Occupational Safety and Health Task Force related to employee-transported contaminants in workers' homes

PL102–375 Older Americans Act
- Study of nursing homes for the elderly
- Study of home health care of the elderly

PL102–381 Department of the Interior and Related Agencies Appropriations Act FY1993
- Study of the Endangered Species Act
- Study of forest product supply

PL102–389 VA–HUD–Independent Agencies Appropriations Act FY1993
- Study of wetlands to identify, measure, and compare wetlands' functions and values

PL102–394 Departments of Labor and of Health and Human Services and Education Appropriations Act FY1993
- Study of the National Institute of Health's Women's Health Initiative

PL102–395 Departments of Commerce and Justice and State Appropriations Act FY1993
- Study of endangered Columbia River salmon stocks

PL102–396 Department of Defense Appropriations
- Study on contingency assumptions and hydrological methodologies used in the Sacramento and American Rivers

PL102–486 Energy Policy Act of 1992
- Study of standards for protection of human health from radioactive materials stored at Yucca Mountain
- Study on energy subsidies
- Study of decontamination and decommissioning of uranium enrichment facilities
- Study to evaluate electric and magnetic field research

PL102–508 Pipeline Safety Improvement Act

- Study on prevention of pollution from hazardous liquid pipelines

PL102–531 Health Promotion and Disease Prevention Act

- Study concerning surgical technique and medical innovation to reduce blood-borne disease transmission

PL102–567 National Oceanic and Atmospheric Administration Authorization Act of 1992

- Study of the science and technology criteria for decisions on closing or reorganizing weather stations
- Study to identify suitable coastal areas for intensive monitoring of water quality
- Study on effects of dolphin feeding

PL102–579 Waste Isolation Pilot Plant (WIPP) Land Withdrawal Act

- Study of an experimental program involving transuranic waste at WIPP

PL102–585 Veterans Health Care Act of 1992

- Study of the health consequences of military service during the first Gulf War

103rd Congress: 25 Mandates

PL103–43 NIH Revitalization Act of 1993

- Study on the adequacy of nurses in hospitals and nursing homes

PL103–66 Budget Reconciliation Act FY1994

- Study of the national vaccine program

PL103–111 Department of Agriculture Appropriations Act

- Study of the scientific basis for listing methyl bromide as a class I controlled substance because of its ozone-depleting potential

PL103–160 Defense Authorization Act (four studies)

- Study of high-performance computing and communications
- Study of global positioning systems—investigation of civil or commercial funding and management for this system
- Study of iodine-131—review of Air Force medical studies conducted on certain Alaskan natives before 1958
- Comprehensive study of cryptography
- Study of cryptographic technologies and national cryptography policy

PL103–206 Coast Guard Authorization Act

- This act expands authority to enter into cooperative agreements between the federal government and the NRC

PL103–227 Goals 2000: Educate America

- Study of the National Goals Panel and National Education Standards and Assessment Council—evaluation of technical work and procedures
- Study of the Bureau of Indian Affairs cost analysis—meeting certain standards of schools funded by the Bureau of Indian Affairs

PL103–272 Transportation, Title 49, US Code, Revision and Enactment

- These are authorizing grants and cooperative agreements with the NAS

PL103–296 Social Security Independence Program Improvements

- Study by the Commission on Childhood Disability in consultation with the NAS on the definition of "disability" as it applies to determining whether children under age 18 are eligible for benefits

PL103–317 Department of Commerce, Justice, State, and the Judiciary Appropriations 1995

- Study of the objectivity, methodology, and applicability of environmental accounting

PL103–322 Omnibus Crime Control and Safe Streets Act

- Study of the control of violence, including rape and domestic violence, against women

PL103–327 Departments of Veterans Affairs, Housing and Urban Development, Independent Agencies Appropriations 1995

- Study of several aspects of research and development activities of the U.S. Environmental Protection Agency

PL103–331 Department of Transportation and Related Agencies Appropriations 1995 (two studies)

- Study of motor vehicle safety labeling
- Study of the National Advanced Driving Simulator

PL103–333 Departments of Labor, Health and Human Services, and Education Appropriations 1995 (two studies)

- Study of federal government research and development activities
- Study of school finance equalization efforts in the United States—a three-year study

PL103–335 Department of Defense Appropriations 1995

- Study on the potential adverse health effects of Army spraying of zinc cadmium sulfide in Minneapolis, St. Louis, and other cities in 1953

PL103–337 Department of Defense Authorization FY1995 (three studies)

- Study of the F-22 aircraft program—determining the desirability of waiving the live-fire survivability tests

- Study of veterans of the first Gulf War—determining the health consequences of serving in southwest Asia
- Study of the domestic production of high-purity electrolytic chromium metal

PL103–382 Improving America's Schools Act of 1994

- Study of the feasibility and validity of assessments and the fairness and accuracy of the data produced by pilot programs

PL103–446 Veteran's Benefits and Improvements Act of 1996 (three studies)

- Study of a medical follow-up agency for oversight and review of research findings of health consequences of service in the Persian Gulf
- Study to evaluate the health consequences for family members of atomic veterans exposed to ionizing radiation
- Yearly review of the adequacy and implementation by the Department of Veterans Affairs of a comprehensive clinical evaluation protocol

104th Congress: 22 Mandates

PL104–88 Interstate Commerce Commission Termination Act 1995

- Study on fiber drum packaging in transportation

PL104–106 Department of Defense Authorization FY1996

- Study of C4I (command and control, communications, computers and intelligence) by the National Research Council

PL104–132 Anti-Terrorism Act

- Study on taggants in fertilizer

PL104–146 Ryan White CARE Reauthorization Act of 1995

- Study of the effectiveness of the states in reducing perinatal HIV (human immunodeficiency virus) transmission and related reduction barriers

PL104–169 National Gambling Impact—Study Commission Act

- Study of the social and economic impacts of gambling in the United States on (1) federal, state, local, and Native American tribal governments and (2) communities and social institutions

PL104–182 Safe Drinking Water Act Amendments of 1996

- Study of the risk for radon in drinking water and an analysis of health risk reduction benefits associated with various mitigation measures to reduce radon levels in indoor air

PL104–193 Personal Responsibility and Work Opportunity Reconciliation Act of 1996

- Study in consultation with the Secretary of Agriculture and the Centers for

Disease Control and Prevention (CDC) on the use of food stamps to purchase vitamins and minerals

PL104–204 Veterans Administration, Housing and Urban Development, Independent Agency Appropriations

- Study on hormone-related toxicants
- Study on EPA's Mobile Source Emissions Factor Model

PL104–208 Department of Defense Appropriations (Omnibus Spending Bill)

- Study on tagging black powder
- Study of the status of research into cancer among minorities and medically underserved populations
- Study on setting up an advisory committee on surface transportation safety
- Study to review and report on the National Biological Survey beginning in 1998 and then every five years

PL104–264 Federal Aviation Administration Research, Engineering, and Development Management Reform Act of 1996

- Study of the effectiveness of weapons and explosive detection systems in commercial aviation

PL104–273 Helium Privatization Act of 1996

- Study on the disposal of helium reserves, including crude helium resources (these reserves will have a substantial adverse effect on U.S. scientific, technical, biomedical, or national security interests)

PL104–297 Sustainable Fisheries Act

- Study on the performance and effectiveness of community developments quota programs under the authority of the North Pacific and Western Pacific Councils

PL104–303 Water Resources Development Act of 1996

- Study on risk analysis and flood control

PL104–324 Coast Guard Authorization Act of 1996

- Study by the Marine Board of the National Research Council of relative environmental and public health risks posed by discharged group-5 fuel oil
- Study by the Marine Board of the National Research Council of tanker vessel lightering

105th Congress: 59 Mandates

PL105–33 Balanced Budget Act of 1997

- Study of the expansion or modification of preventive or other benefits provided to Medicare beneficiaries under Title XVIII of the Social Security Act

- Study of payments for clinical laboratory tests under Title XVIII Part B of the Social Security Act
- Study of the accessibility and quality of health care for low-income individuals who are a consideration in the quality-assurance programs and accreditation standards applicable to managed care entities in the private sector
- Study of early and periodic screening, diagnostic, and treatment services

PL105–65 Departments of Veterans Affairs and Housing and Urban Development and Independent Agencies Appropriations Act of 1998

- Study to develop a comprehensive, prioritized, near- and long-term program for particulate matter research and monitoring
- Study to evaluate the engineering challenges posed by extravehicular-activity requirements of space station construction and assembly
- Study of the effectiveness of the Environmental Protection Agency's inspection and maintenance programs
- Study of the availability, effectiveness, costs, and effects of technologies for the remediation of sediments contaminated with polychlorinated biphenyls, including dredging and disposal

PL105–66 Department of Transportation and Related Agencies Appropriations Act of 1998

- Study of the research and development plans and programs of the Federal Railroad Administration

PL105–78 Departments of Labor, Health and Human Services, and Education Appropriations Act of 1998

- Study of the policies and process used by the National Institutes of Health to determine funding allocations for biomedical research
- Study to determine if an equivalency scale can be developed that would allow scores from commercial standardized tests to be compared with state assessments and with the National Assessment of Educational Progress
- Study of test items developed or funded by a federal agency to determine their quality, validity, adequacy, and freedom from bias for tracking the graduation or promotion of students
- Study and recommendations for appropriate methods, practices, and safeguards to ensure that tests to assess student performance are accurate and not discriminatory
- Study of the effects of televised alcohol advertising on youth alcohol consumption
- Study of welfare reform outcomes

PL105–85 National Defense Authorization Act FY1998

- Study of aspects of the chemical non-stockpile materiel project

PL105–86 Agriculture, Rural Development, Food and Drug Administration, and Related Agencies Appropriations Act of 1998

- Study on the scientific and organizational needs for an effective food safety system

PL105–115 Food and Drug Administration Modernization Act of 1997

- Study of the effect on humans of the use of elemental, organic, or inorganic mercury as a drug or dietary supplement
- Study of the scientific issues raised as a result of an amendment to the Federal Food, Drug, and Cosmetic Act and the Food Health Service Act

PL105–119 Departments of Commerce, Justice, and State, the Judiciary and Related Agencies Appropriations Act of 1998

- Study of the most effective techniques and technologies to block children from receiving pornographic images via the Internet
- Study of the role of marine sanctuaries in marine resource conservation, as well as the usefulness of marine reserves, including their effects on water quality and the abundance of living marine resources
- Study of fisheries resources of summer flounder

PL105–178 Transportation Equity Act for the 21st Century

- Study of the congestion mitigation and air quality improvement program
- Study to consult on the production of a documentary on infrastructure awareness
- Study of the regulation of weights, lengths, and widths of commercial motor vehicles operating on highways that receive federal aid to which federal regulations apply
- Study of the effect of contracting out mass transportation operation and administrative functions
- Study of the safety issues attendant to the transportation of schoolchildren to and from school and school-related activities by various transportation modes
- Study of the integrated surface transportation research and technology development strategic plan
- Study to determine the goals, purposes, research agenda and projects, administrative structure, and fiscal needs for a new strategic highway research program
- Study of statistical quality standards regarding accuracy and reliability measures

PL105–185 Agricultural Research, Extension, and Education Reform Act of 1998

- Study to consult on the administration of the Agricultural Genome Initiative
- Study of the role and mission of federally funded agricultural research, extension, and education

PL105–261 National Defense Authorization Act FY1999

- Study to consult on determinations and certifications applying to an alternative technology for the destruction of lethal chemical munitions, other than incineration
- Study on the technology base of the Department of Defense

PL105–276 Departments of Veterans Affairs and Housing and Urban Development and Independent Agencies Appropriations Act 1999

- Study of the potential toxicologic risks of all flame-retardant chemicals identified as likely candidates for use in residential upholstered furniture
- Study to develop methods for evaluating federally funded research and development programs
- Study of the Environmental Protection Agency's Clean Air Science Advisory Committee with respect to the speciation component of the agency's particulate matter monitoring plan
- Study of the effects of copper in drinking water on human health
- Health research to prepare recommendations on an appropriate mercury exposure reference dose

PL105–277 Omnibus Consolidated and Emergency Supplemental Appropriations FY1999

- Study of the environmental and reclamation requirements related to mining of locatable minerals on federal lands
- Study of the available scientific literature examining the cause-and-effect relationship between repetitive tasks in the workplace and musculoskeletal disorders
- Review of the current policies of the Organ Procurement and Transplantation Network
- Study of the technical feasibility, validity, and reliability of including test items from the National Assessment of Educational Progress for fourth-grade reading and eighth-grade mathematics
- Update of the 1991 NAS study of the airline industry and deregulation
- Study of the status of older workers in the information technology field
- Study on the available scientific evidence regarding associations between illnesses and exposure to toxic agents, environmental or wartime hazards, or preventive medicines or vaccines associated with service in the first Gulf War
- Study of the appropriate amounts of fruit, fiber, and sugar in the diet of the population, targeted for benefit by the Special Supplemental Nutrition Program for Women, Infants, and Children
- Study to establish and coordinate a national task force on alcohol-related birth defects
- Study of the current scientific knowledge of the potential environmental causes of breast cancer
- Study of the Gang Resistance Education and Training program

PL105–301 Crime Victims with Disabilities Awareness Act

- A study to increase knowledge and information about crimes against individuals with developmental disabilities

PL105–305 Next Generation Internet Research Act of 1998

- Study of the short-term and long-term effects on trademark rights of adding new generic top-level domains and related dispute resolution procedures

PL105–314 Protection of Children from Sexual Predators Act of 1998

- Study of computer-based technologies and other approaches to the problem of the availability of pornographic material to children on the Internet

PL105–368 Veterans Programs Enhancement Act of 1998

- Review and evaluation of the available scientific evidence regarding associations between illness and service in the first Gulf War
- Study developing a plan for the establishment of a national center or national centers for the study of war-related illnesses and postdeployment health issues
- Study to assess whether a methodology could be used for determining the efficacy of treatments furnished to and health outcomes of veterans of the first Gulf War treated for illnesses associated with their service
- Study to develop a curriculum for the care and treatment veterans of the first Gulf War who have undiagnosed illnesses and periodically review and provide recommendations regarding research plans and research strategies

PL105–383 Coast Guard Authorization Act of 1998

- Study to establish a rationally based equivalency assessment approach that accounts for the overall environmental performance of alternative tank vessel designs

PL105–392 Health Professions Education Partnerships Act of 1998

- Study on the training needs of health professionals with respect to the detection and referral of victims of family or acquaintance violence

106th Congress: 45 Mandates

PL106–69 Department of Transportation and Related Agencies Appropriations Act 2000

- Study on privatizing the traffic management function within the Federal Aviation Administration

PL106–71 Missing, Exploited, and Runaway Children Protection Act

- Study on the antecedents of school violence in urban, suburban, and rural schools

PL106–74 Departments of Veterans Affairs and Housing and Urban Development and Independent Agencies Appropriations Act 2000

- Study of the effectiveness of clean air programs used by federal, state, and local governments
- Cumulative effects study of North Slope oil and gas development
- Study of the availability and usefulness of data collected from all of NASA's science missions
- Evaluation of the scientific goals of the Triana mission

PL106–79 Department of Defense Appropriations Act 2000

- Study of the effectiveness and safety of the anthrax vaccine

PL106–113 An Act Making Consolidated Appropriations for the Fiscal Year Ending September 30, 2000, and for Other Purposes

- Retrospective study of the costs and benefits of federal research and development technologies in the areas of fossil energy and energy efficiency
- Study of the Occupational Safety and Health Administration's proposed rule relating to occupational exposure to tuberculosis
- Study on ethnic bias in medicine
- Study on further language regarding the study on school violence from Public Law 106–71

PL106–129 Healthcare Research and Quality Act of 1999

- Review of quality oversight, quality improvement, and quality research programs in the health services

PL106–181 Wendell H. Ford Aviation Investment and Reform Act for the 21st Century

- Study of air quality in passenger cabins of aircraft used in domestic and foreign air transportation

PL106–193 Methane Hydrate Research and Development Act of 2000

- Study of the progress made under the methane hydrate research and development program and recommendations for future needs

PL106–291 Department of the Interior and Related Agencies Appropriations Act 2001

- Study of Atlantic salmon
- Study of the potential of homogenous charge combustion ignition technology in the next annual review of the PNGV (Partnership for the New Generation of Vehicles) program

PL106–310 Children's Health Act of 2000

- Study to determine the costs of immunosuppressive drugs provided to children pursuant to organ transplants and the extent to which health plans and health insurance cover such costs

- Study of issues related to the treatment of phenylketonuria and other metabolic disorders and mechanisms to ensure access to effective treatment, including special diets
- Study on the development of medications for the treatment of addiction to amphetamine and methamphetamine

PL106–345 Ryan White CARE Act Amendments of 2000

- Study on reducing the incidence of perinatal transmission of HIV, including the barriers to testing
- Study of the surveillance systems of states on the prevalence of HIV
- Study of the appropriate epidemiological measures and their relationship to the financing and delivery of primary care and health-related support services for low-income, uninsured, and underinsured individuals with HIV disease

PL106–346 Department of Transportation and Related Agencies Appropriations Act of 2001

- Study on whether the static stability factor is a scientifically valid measurement in comparison to dynamic testing
- Evaluation of the effectiveness and effects of CAFE (corporate average fuel economy) standards

PL106–377 Departments of Veterans Affairs and Housing and Urban Development and Independent Agencies Appropriations Act of 2001

- Study of carbon monoxide episodes in meteorological and topographical problem areas
- Review of the quality of science used to develop and implement TMDLs (total maximum daily loads)

PL106–387 Agriculture, Rural Development, Food and Drug Administration, and Related Agencies Appropriations Act of 2001

- Evaluation of the role of scientifically determined criteria in the production and regulation of meat and poultry products

PL106–391 National Aeronautics and Space Administration Authorization Act of 2000

- Jointly with the National Academy of Public Administration, study of the status of life and microgravity research as it relates to the International Space Station
- Review of efforts to determine the extent of life in the universe

PL106–398 National Defense Authorization Act FY2001

- Evaluation of the costs, benefits, and risks associated with various remediation alternatives for the cleanup of the Moab uranium milling site

PL106–419 Veterans Benefits and Health Care Improvement Act of 2000

- Review of the dose reconstruction program of the Defense Threat Reduction Agency

PL106–525 Minority Health and Health Disparities Research and Education Act of 2000

- Study of the Department of Health and Human Services' data collection systems and practices relating to the collection of data on race or ethnicity

PL106–528 Airport Security Improvement Act of 2000

- Study of airport noise

PL106–541 Water Resources Development Act of 2000

- Study of the independent peer review of feasibility reports for water resources projects
- Review of the methods used to conduct economic and environmental analyses of water resources projects
- Biennial review of the progress of the Comprehensive Everglades Restoration Plan

PL106–554 Consolidated Appropriations Act 2001

- Study of coal waste impoundments and alternatives for future coal waste disposal
- Study to evaluate the effectiveness of the current role and structure of the Ryan White CARE Act and the efforts to create a national consumer and provider education center within the pediatric HIV–AIDS community
- Study of the structure of the National Institutes of Health
- Study to evaluate risk assessment in relation to children's health and safety
- Study of the consequences of high-stakes testing
- Pursuant to HR2090, as passed by the House of Representatives on September 12, 2000, study of the feasibility and social value of a coordinated oceanography program
- Independent scientific review of the November 30, 2000, Biological Opinion for the Bering Sea–Aleutian Islands and Gulf of Alaska groundfish fisheries
- Study on Medicare coverage of routine thyroid screening
- Comprehensive study that provides semiannual progress reports on how the SBIR (Small Business Innovation Research) program has stimulated technological innovation and used small businesses to meet federal research and development needs

Section 15 of the Federal Advisory Committee Act as Added by Public Law 105–153

(a) IN GENERAL. An agency may not use any advice or recommendation provided by the National Academy of Sciences or National Academy of Public Administration that was developed by use of a committee created by that academy under an agreement with an agency, unless—

(1) the committee was not subject to any actual management or control by an agency or an officer of the Federal Government;

(2) in the case of a committee created after the date of enactment of the Federal Advisory Committee Act Amendments of 1997, the membership of the committee was appointed in accordance with the requirements described in subsection (b)(1); and
(3) in developing the advice or recommendation, the academy complied with—
    (A) subsection (b)(2) through (6), in the case of any advice or recommendation provided by the National Academy of Sciences; or
    (B) subsection (b)(2) and (5), in the case of any advice or recommendation provided by the National Academy of Public Administration.

(b) REQUIREMENTS. The requirements referred to in subsection (a) are as follows:
(1) The Academy shall determine and provide public notice of the names and brief biographies of individuals that the Academy appoints or intends to appoint to serve on the committee. The Academy shall determine and provide a reasonable opportunity for the public to comment on such appointments before they are made or, if the Academy determines such prior comment is not practicable, in the period immediately following the appointments. The Academy shall make its best efforts to ensure that (A) no individual appointed to serve on the committee has a conflict of interest that is relevant to the functions to be performed, unless such conflict is promptly and publicly disclosed and the Academy determines that the conflict is unavoidable, (B) the committee membership is fairly balanced as determined by the Academy to be appropriate for the functions to be performed, and (C) the final report of the Academy will be the result of the Academy's independent judgment. The Academy shall require that individuals that the Academy appoints or intends to appoint to serve on the committee inform the Academy of the individual's conflicts of interest that are relevant to the functions to be performed.
(2) The Academy shall determine and provide public notice of committee meetings that will be open to the public.
(3) The Academy shall ensure that meetings of the committee to gather data from individuals who are not officials, agents, or employees of the Academy are open to the public, unless the Academy determines that a meeting would disclose matters described in section 552(b) of title 5, United States Code. The Academy shall make available to the public, at reasonable charge if appropriate, written materials presented to the committee by individuals who are not officials, agents, or employees of the Academy, unless the Academy determines that making material available would disclose matters described in that section.
(4) The Academy shall make available to the public as soon as practicable, at reasonable charge if appropriate, a brief summary of any committee meeting that is not a data gathering meeting, unless the Academy determines that the summary would disclose matters described in section 552(b) of title 5, United States Code. The summary shall identify the committee

members present, the topics discussed, materials made available to the committee, and such other matters that the Academy determines should be included.

(5) The Academy shall make available to the public its final report, at reasonable charge if appropriate, unless the Academy determines that the report would disclose matters described in section 552(b) of title 5, United States Code. If the Academy determines that the report would disclose matters described in that section, the Academy shall make public an abbreviated version of the report that does not disclose those matters.

(6) After publication of the final report, the Academy shall make publicly available the names of the principal reviewers who reviewed the report in draft form and who are not officials, agents, or employees of the Academy.

(c) REGULATIONS. The Administrator of General Services may issue regulations implementing this section.

Note

[1] Two government organizations go by the initials NRC: the National Research Council, created in 1916, and the U.S. Nuclear Regulatory Commission, established after the reorganization of the U.S. Atomic Energy Commission in 1976.

Appendix 3

An External Evaluation of the GAO's First Pilot Technology Assessment

Note: This appendix reproduces the main body of a report prepared by Fri et al. (2002) that evaluated the first pilot technology assessment performed by the General Accounting Office (GAO) on using biometrics for border security. The format of the report has been slightly edited, two figures have been dropped, and appendices that appear in the original are not included.

The original GAO report (GAO 2002) can also be found in Portable Document Format (PDF) at GAO's website (www.gao.gov) by searching for report GAO–03–174 or at the U.S. Government Printing Office website by searching on the report number at www.gpoaccess.gov/gaoreports/.

Abstract

In response to a request from the Congress in late 2001, the GAO has performed a technology assessment of biometric technology for border control. In addition to being deeply interested in the substance of this topic, the Congressional requesters have also viewed this as a pilot experiment to explore whether, in addition to its more traditional functions of performance assessment and auditing, the GAO might be able to provide the Congress with support in technology assessment. Our three-person committee was created to provide an independent external evaluation of this effort.

Given the relatively short time that was available to perform the assessment, and the fact that GAO has not previously engaged in such analysis, they have done a very good job. They successfully identified, described, and evaluated a range of relevant technologies. With assistance from the NRC, they also engaged, and in their report have reflected, the views held by experts, affected parties, and concerned stakeholders so that Congressional Staff and Members can accurately identify and weigh the tradeoffs involved in policy choices.

This first GAO effort has been rather less successful in framing the analysis broadly in such a way as to address the full range of issues which the Congress is likely to have to consider and in providing analytically informed input that will support the needs of Congressional staff and Members as they refine and tune legislative products. These limitations result from the fact that while GAO has staff with excellent technical credentials, they have relatively little previous experience in framing and performing policy analysis. In addition, current internal GAO guidelines are not entirely consistent with the needs of technology-focused policy analysis and assessment.

While it will pose a significant challenge, we believe that if the Congress chooses to ask the GAO to perform additional technology assessments in the future, the problems we have identified can be resolved if GAO acquires more policy–analytic staff capabilities, makes appropriate changes in internal guidance and administrative arrangements, and makes effective use of outside expertise and contractors.

Introduction

When the GAO began its assessment of Technologies for Border Control, they decided to simultaneously arrange to have an evaluation performed of the process they used and the assessment they produced. Accordingly, they constituted the three of us as a small independent committee and charged us with performing an evaluation of their work as it progressed.

During the course of the assessment, we met several times with Naba Barkakati and others at GAO, with a number of Congressional staff, with staff at the National Research Council, and with others. We experienced excellent cooperation from all those with whom we met. (*Editor's note:* A full list of the people with whom discussions were held was included as an attachment to the original report.)

In performing this evaluation, we felt it important to explore and attempt to answer four questions:

1. What was the process by which the assessment was commissioned and the topic and scope of the study defined?
2. What is the quality of the final product?
3. What was the process used by the GAO in performing this assessment? How was it different from GAO's standard audit process?
4. If GAO were asked do more assessments, how could their capacity to do so be strengthened? What should Congress hope to get if and when it seeks future technology assessments, and, can the GAO meet that need?

We address each of these questions in turn in the four numbered sections that follow.

1. Commissioning the Assessment and Defining the Topic

In June of 2001, Sen. Jeff Bingaman (D-NM) asked his staff to investigate the possibility of having one of the Congressional support agencies conduct a pilot technology assessment. His goal was to see whether there was a way to provide the Legislative Branch with a technology assessment capability without the large investment of funding and people that the old Office of Technology Assessment (OTA) had required.

Sen. Bingaman's Legislative Assistant Jonathan Epstein held discussions with the Congressional support agencies at the Library of Congress/Congressional Research Service (CRS) and the General Accounting Office (GAO) to determine their willingness to take on the pilot assessment program. GAO was amenable to doing the work and was seen by staff as better suited to take on this task than CRS. On July 19, 2001, an amendment was added to the Senate version of the Fiscal Year 2002 Legislative Branch Appropriations Bill, S 1172, to fund a $1-million technology assessment pilot program. The comptroller's office in GAO was told of this action four days later by e-mail and given some guidance on the intent of Sen. Bingaman's amendment. The National Research Council's Office of Congressional Affairs was copied on this message.

There was no similar provision in the House version of the Fiscal Year 2002 Legislative Branch Appropriations Bill, HR 2647. In conference with the House of Representatives on the legislation, the technology assessment funding was cut to $500,000, and the conference report instructed GAO's Comptroller General to, "... obligate up to $500,000 of the funds made available for a pilot program in technology assessment as determined by the Senate and to submit to the Senate a report on the pilot program not later than June 15, 2002." The language indicates little interest on the part of the House Appropriations Committee in the pilot technology assessment, although House Science Committee Members and staff were interested and subsequently the Committee sent a support letter.

On November 12, 2001, the FY 2002 Legislative Branch Appropriations Bill was signed into law (PL 107-68), and two days later, the comptroller's office in GAO was again contacted and informed of this action. The office was told that a subject for the pilot assessment would be forthcoming from Congressional Staff by the week of December 10, 2001. A series of meetings among Congressional staff (and consultation with National Academy of Sciences staff) resulted in a broad topic area being selected and on November 30, 2001, a draft Congressional request letter was shared with the comptroller's office in GAO with the assessment topic noted as "The Future of US Border Control—The Role of Technology." The formal request letter was sent to GAO from the four Senate requestors on December 17, 2001 (a letter of interest in the pilot assessment program was sent on December 10, 2001, by six House members).

For the next few weeks, Congressional Staff and GAO staff attempted to narrow the scope of the request, given the deadline for delivery of the assessment by June 15, 2002. Several weeks were lost because GAO initially assigned an audit team familiar with the Immigration and Naturalization Service to the task, not understanding that the request was for a technology assessment and not a standard GAO audit. After some dialogue between GAO and Congressional Staff, the narrowing of the topic to "biometrics" was accomplished and a work plan was put in place.

GAO contacted NAS staff in early February 2002 and formalized a request for National Research Council (NRC) assistance on February 25 under a standing contractual arrangement that had been recently negotiated (see discussion below in Section 3.2). The GAO Task order (GAO 4713-103) on "Assessing the Maturity of Biometric Technologies for Use in U.S. Border Control and Their Policy Implications" called on the NRC to conduct a workshop of experts and stakeholders. It was subsequently decided that two workshops were required. The first, held on April 25 and 26, 2002, focused mainly on the state and cost of the technology. The second, held on May 16 and 17, 2002, examined privacy concerns and policy implications.

Previously, on October 26, 2001, the USA Patriot Act was signed into law (PL 107-56) with a provision requiring the National Institute of Standards and Technology (NIST) to develop a technology standard within two years to provide a standard for, "a cross-agency, cross-platform electronic system that is a cost-effective, efficient, fully integrated means to share law enforcement and intelligence information necessary to confirm the identity of such persons applying for a United States visa or such persons seeking to enter the United States pursuant to a visa." This act elevated the importance of the pilot GAO assessment when it was decided to focus it in this area of technology.

On May 14, 2002, the Border Security Act became law (PL 107-173), with a requirement that by October 26, 2004, the U.S. would have installed at all ports of entry, "... equipment and software to allow biometric comparison and authentication of all United States visas and other travel and entry documents issued to aliens." This Act made the findings of the pilot assessment even more critical to the policy process, especially given the Senate's having moved the deadline back a year because of the uncertainties surrounding the technology.

With the delays in getting the assessment funded, focused, and started, the original deadline for delivery of the report by June 15, 2002, was changed to September 2002. In lieu of a report, the Hill staff were briefed by GAO on the status and initial findings in a series of briefings in June 2002. The House and Senate authorizing committee and Member staff were briefed on June 6, 2002, and Senate Appropriations Committee staff were briefed on June 10, 2002. GAO staff held a set of subsequent briefings to Hill staff on the draft assessment summary. Senate Appropriations Committee staff were briefed on July 29, 2002, and the authorizing committee and Member staff were briefed on July 31, 2002.

2. Our Assessment of the Assessment

What should a successful technology assessment, performed for the U.S. Congress, accomplish? While the three of us have significant prior experience in this area, we decided that we should not answer this question ourselves but should seek the views of a group of experienced Congressional staff. They suggested that a successful assessment should

- be framed in such a way as to address a problem of active concern to the Congress;
- produce results in a timely manner;
- be scientifically complete and credible;
- be balanced in the way it frames, analyzes, and discusses the issues;
- reflect the full range of views held by experts and affected parties and concerned stakeholders so that staff and Members can accurately identify and weigh the tradeoffs involved in policy choices; and
- provide analytically informed input that supports the needs of Congressional staff and Members as they refine and tune legislative products.

These criteria correspond closely with our own views of what is needed.

Staff told us that in addition to the assessment product itself, it was important to them to have ongoing access to members of the assessment team, and the knowledge they have acquired on the topic, since they often need iterative interaction as they refine and tune legislative products or engage in other work related to the topic of the assessment.

The current GAO report does a good job on some of these objectives and falls somewhat short on others. At the outset, it must be acknowledged that the GAO staff produced a quality, detailed report on a very short deadline, especially given the GAO procedural changes required to conduct this pilot assessment, as compared to a standard GAO audit. The delays in enactment of the funding and in selecting the topic for this assessment exacerbated the time factor. Both the GAO and Congressional Staff indicate that communications with the other were very good once the report topic was selected and work commenced.

The GAO sought to include a broad range of views. This is evident in the range of people invited to the National Academy of Sciences (NAS) Workshops, which included a range of agency and technical experts, as well as several non-technical groups in the second workshop. GAO staff were open to comments and suggestions made by NAS staff on the focus and conduct of both of those workshops. They continued to invite non-federal involvement with the issuance of review drafts of the report to some of the outside workshop participants.

The final report does a very good job of summarizing the processes of border entry and examining the basic problem of positive identification at border entry points. Through site visits and communications with implementing agencies and departments (State Department and Immigration and Naturalization Service), the assessment team gained a good operational understanding of the visa and passport issuance and entry process that is reflected in the report. With contact with technology providers, both at the workshops held at the NAS and through individual contacts, the assessment team assembled a comprehensive view of currently available biometric technologies. A variety of factual details and errors were successfully clarified through the review process.

The analysis conducted using these operational understandings together with the knowledge of the currently available biometric technologies resulted in a good review of the inherent strengths and weaknesses of the technologies as applied in real-life situations. The use of the four scenarios to further test assumptions was an effective way to impose rigor and relevancy on the analyses (note, however, that as of last May, Congress has effectively already opted for Scenario 3).

The cost analysis performed was solid and provides to policymakers the first good estimate of what resources will be required to apply these technologies, although we suspect that broader system requirements and costs may, in some cases, have been underestimated. The building of the analytical base through the understanding of the technology and the operational realities of the affected agencies was critical, but so was the blending of this work with the more traditional GAO audit/cost accounting function that resulted in these cost estimates.

The report is rather less successful in framing and addressing the broad policy issues that technology for personal identification would be used to address, placing the evaluation of specific technologies in a broader system context, and providing comparative insight with respect to other, more conventional options.

While early drafts of the report completely overlooked these issues, they were identified as important through higher level internal review at GAO and by a number of external reviewers. In both the opening section, titled "The Technology Assessment at a Glance" and in the Chapter 6 Summary, the final version of the report does clearly recognize that biometric technology must be viewed as one part of a much larger complex problem of border controls. And, it appropriately concludes that decisions about whether and how to deploy biometric technology should be based on a broad analysis of that system—an analysis that the GAO report refers to as a high level "risk based approach" that uses "cost–benefit analysis."

We recognize that a full review of these broader policy issues in a system context, and a full comparative analysis of more conventional options, was not feasible given the time and resource constraints imposed upon this process. However, included in the report are sufficient details and data that would allow a better focused discussion of some of these points.

For example, the assessment notes that "the desired benefit of all of the described scenarios—the use of biometric watch lists or biometrically enabled travel documents—is to prevent the entry of travelers who are inadmissible into the United States." If this is the objective, the potential effectiveness of the technology can immediately be bounded. The report tells us that there are about 275,000 illegal entries per year, of whom approximately 60% sneak in across the open borders without a visa and without encountering a port of entry. Thus, at best, biometric technologies, perfectly applied at border crossings today, could eliminate about 40% of the problem (a fact that is noted in the final report but was not noted in early internal drafts). This fraction might be even lower if effective biometric technology induced larger numbers of people to adopt the route of illegal entry.

These considerations suggest a reframing of the assessment question in terms of overall system-wide cost effectiveness. If the objective is "to prevent the entry of travelers who are inadmissible into the United States," what is the most cost-effective way to do that, and what, if any, role should biometric technologies play in achieving that goal?

If we limit ourselves to a narrower consideration of controlling the entry of only those persons who pass through ports of entry, a broader system perspective is still essential if the analysis is to provide the insight that a decision-maker will need to decide whether a given level of performance improvement could best be achieved through the use of a biometric-based system or through enhanced application of more conventional methods—including better use of conventional information technologies.

Consider, for example, the scenario of issuing visas with biometrics. The assessment estimates that the initial cost of implementing this system would be "between $1.3 and $2.9 billion" and that the annual recurring costs would be "between $0.7 and $1.5 billion." While this is certainly useful information, the more pertinent policy questions are:

- how large a reduction in inadmissible entry are these investments likely to achieve?
- could a similar (or larger) reduction in inadmissible entries be achieved using more conventional methods, and if so, how would the cost for such a conventional system compare with the cost of the biometrics system—would it be smaller or larger? Would there likely be ancillary benefits (such as skill or alertness levels of INS staff) associated with one or the other approach that are not reflected in the costs?

In order to answer such questions, the assessment team would have had to build a simple simulation model of the current visa system which included estimates of the performance characteristics of that system, built a similar model for the hypothetical biometric system, and then performed systematic comparative analysis. Such an analysis would certainly have been possible in the time available, but, anticipating its need and implementing it effectively would have required a broad policy-sensitive perspective and a high level of policy–analytic experience.

To continue this example one more step, one of the primary objectives of a biometrics visa system would be to minimize the use of fraudulent visas. It would have been useful if the report had been able to estimate the cost of creating fraudulent visas under the current system and the proposed biometrics system. Clearly the cost of forging a visa for a biometric system would be higher. How much higher? Would rogue nations and large terrorist organizations be able to afford the technology to produce such forgeries? If so, how much of the national policy objective would be achieved if, while large numbers of "little fish" were prevented from fraudulently entering, "big fish" could still afford to create forgeries. The report briefly notes (in its Table 11) that modern encryption technology might be used to reduce or eliminate the possibility of certain kinds of forgery. While true, the use of such technology raises complex system-level issues which add technical and managerial complications. These system issues have not received much consideration in the current report, but they could be crucial in determining the effectiveness of the system in protecting against the most dangerous inadmissible persons. They are also the basis for our suggestion above that some of the costs may be underestimated.

There are several other issues that might have been addressed in order to help the Congress think about how best "to prevent the entry of travelers who are inadmissible into the United States." For example:

- The assessment notes that some fraction of those who enter legally overstay their visas but observes that little is done to track them down. It would be interesting to consider whether biometric technology could play a cost-effective role in expanding that effort.
- The assessment notes that 60% of illegal entries involve people sneaking across borders. The U.S. Immigration and Naturalization Service (INS) may have some idea of the marginal cost of preventing one more entry via stepped up border security and patrols (they certainly know the average cost of intercepted entries). How does that cost compare with the marginal (or average) cost of a biometric system, assuming that it could perfectly intercept all of the approximately 165,000 inadmissible persons each year at border crossings? What happens to these costs if one assumes more realistic detection rates?
- If 60% of illegal entries are across our borders, what would be the cost-effectiveness of investments in helping the Canadians and Mexicans improve their visa and entry control systems compare with investments in improving the U.S. system? This issue becomes more pertinent as NAFTA (the North American Free Trade Agreement) opens our borders with those countries—as noted by the GAO report in its discussion of how few people entering the U.S. have to carry visas.

Finally, there is the question:

- Is there *any* level of investment in biometrics technology that can meet current statutory deadlines?

Table A3-1. Evaluation of the GAO Technology Assessment in Terms of the Criteria Outlined in Section 2 of the Report

Criteria	*Performance*
Framed in such a way as to address a problem of active concern to the Congress	Limited success
Produce results in a timely manner	Largely successful
Scientifically complete and credible	Successful in the areas covered
Be balanced in the way it frames, analyzes and discusses the issues	Balanced in what was addressed but the focus was too narrow
Reflect the full range of views held by experts and affected parties and concerned stakeholders so that staff and Members can accurately identify and weigh the tradeoffs involved in policy choices	Largely successful
Provide analytically informed input that supports the needs of Congressional staff and Members as they refine and tune legislative products	Limited success

The assessment makes only one set of references to those deadlines in the context of its discussion of the federally mandated Chimera system. This section outlines the legal requirement for system implementation by October 26, 2004, the requirement for NIST standard setting by October 26, 2003, and the requirement that privacy concerns be addressed within two months—by October 26, 2002. However, we found no discussion of these deadlines or the feasibility of meeting them at any other place in the document. The discussion of standards does not mention the mandate to NIST, and the discussion of privacy concerns makes no mention of the upcoming requirements on privacy issues.

None of these criticisms should be taken as minimizing the very considerable accomplishments that the GAO staff have made in a very short period of time. But, as we discuss below, the issues we have raised do illustrate the need to build considerably more policy–analytic capability at the GAO (or through GAO supervised contractors), and to revise the methods the GAO uses for review and external consultation, if the Congress wishes to continue to use this mechanism as a source of technology assessment and analytical insight in the future.

In terms of the criteria outlined at the beginning of this section, we rate the GAO's performance as indicated in Table A3-1.

As elaborated below, we believe that all of the problems encountered could be adequately resolved in future assessments if GAO acquires more policy–analytic staff capabilities, makes appropriate changes in internal guidance and administrative arrangements, and makes effective use of outside expertise and contractors.

3. How the Assessment was Performed

As with any organization, the GAO has a number of firmly established standard operating procedures. To a much greater extent than many organizations, the GAO has carefully documented these procedures, which are collectively referred to as the GAO engagement management process.

3.1. The GAO Engagement Management Process and How it Was Modified for this Assessment

The GAO conducts a large number of financial, program, and management audits and program evaluations each year. Its ability to conduct these projects with the highest level of professional competence is a core strength of the organization. To preserve and enhance this strength, GAO has developed a disciplined process for managing each of its engagements. GAO staff use this seven-step process (*Editor's note:* A figure in the original has been excluded here in this appendix.) as a template for all GAO projects.

The typical GAO project relies on careful fact-finding as a basis for assessing whether the subject of the engagement meets a specified standard of performance. The standard may, for example, be generally accepted accounting principles, legislative or regulatory requirements, or best management practices. The GAO engagement management process is adapted to this analytic paradigm and, based on our cursory inspection, certainly seems to make good sense in that context.

Technology assessment does not, however, rest on the same paradigm. Careful fact-finding is, of course, vital to technology assessment. What is different is what one does with the facts. The aim of technology assessment is to elucidate what might be, to explore the consequences of deploying, applying, and managing a new technology in a variety of possible ways, and the associated implications for policy makers. This requires policy analysis rather than comparison to an established standard.

The biometrics technology assessment used the GAO engagement management template as a guide. Because of the differences in analytic paradigms outlined above, however, it was necessary to modify the standard template in several important ways. The lessons from this experience should be considered if GAO conducts additional technology assessments.

Defining the engagement objective. Defining the scope of this assessment project required considerable iteration with the client. The assessment was originally defined more broadly in terms of technology and border control. Only after several discussions was the final objective narrowed to a focus on biometrics to provide a scope that was both tractable for the analyst (especially given the time frame involved) and useful to the Congress.

Although the circumstances surrounding the initiation of the biometrics project were unique, our experience suggests that similar iteration in choosing and refining the problem definition and scope occurs on virtually all technology assessment projects. However, the GAO template does not nominally provide for substantial client contact until the third "Gate" in its process. While such contact is not prevented, the process for a technology assessment project should require this contact as early as Gate 1. Early attention to defining the objective is important for ensuring that the client's expectations can be met, as well as for providing GAO management the information they need to determine whether the project can be accepted and, if so, what risk level to assign it.

Designing and staffing the project. The GAO engagement management template provides for staffing the project chiefly from internal GAO resources or stakeholders. Ordinarily, the project would be led by the program team most familiar with the agency or program being reviewed. As noted in Section 1, that was initially done in this case as well, but because of the substantial technical content of the task at hand, the project was subsequently reassigned to the Applied Research and Methods (ARM) group. This group staffed the project with its own people and with experts drawn from the IT group, and then added internal legal, economics, and statistics support. In addition, as explained below, the biometrics team recognized the need for input from external technical expertise on biometric technologies and made use of GAO's arrangement with the NRC to provide it (see discussion below in Section 3.2).

While the ARM team managed to bring together considerable expertise in a relatively short time, three problems arose:

- The need for policy expertise in a technology assessment was not fully anticipated. In the case of the biometrics project, policy questions initially centered on privacy issues but evolved to include questions of economic impact on border communities, technology standards, and others. This expertise was not available within GAO, and the biometrics team extended the NRC support to hold a second workshop focused on policy issues.
- As the technology assessment unfolded, knowledge of the overall border control process proved to be increasingly important. Indeed, the need for a risk assessment of the border control process became a major recommendation as the report was written and evaluated. This evolution suggests to us that earlier access to experienced policy analytic staff would have benefited the assessment.
- The ARM team conducts relatively few of its own projects. It successfully managed this project but lacked some of the support resources usually made available. In particular, it did not have its own support for the documentation process required of GAO engagements.

Conducting the analysis. In our opinion, the strengths and limits of the GAO engagement management process came most clearly into focus during the analytic phase of the biometrics project. This work can be divided into three major parts:

- The biometrics team engaged in extensive fact-finding. This step included the NRC panels, interviews with INS and State, and in-depth characterization of four key biometric technologies. These data were documented and vetted in accordance with standard GAO procedures.
- The team created several scenarios as a basis for examining how biometrics technology could be used in border control and visa applications. Initially, these scenarios were devised as context for doing the cost analysis required in the request for the technology assessment. As the project proceeded, however, these scenarios played a larger analytic role in evaluating the effectiveness and overall implications of biometrics technology.
- Finally, the team developed conclusions from the fact-finding and analysis. What emerged as the overarching conclusions were that biometrics technology is not itself an answer to border control problems, and that a high-level risk analysis of border control would be needed to establish the context for making decisions about deploying biometric technology.

The GAO engagement management template focuses almost entirely on the fact-finding step. Guidance on analytic activity is contained in Gate 4 of the template, and our review of the supporting materials for that gate shows that they deal chiefly with the quality of evidence elicited in the fact-finding activity. This same scope is reflected in the Gate 3 step suggesting the use of a design matrix for organizing the analytic effort on an engagement. As a basis for solid fact-finding, the guidance given to engagement teams is excellent, and its use, in our opinion, was appropriate for this phase of the technology assessment.

In contrast, the scenario analysis is not fact-finding but does form an essential part of the foundation on which this technology assessment rests. Its validity cannot be tested in the same way as fact-finding because the analysis is a construct, not an observation. More important, its direction evolved over the course of the project, leading (as is always the case in analyses of this sort) to the possibility that preconceptions are shaping the assumptions and methodologies being used. Nothing in the GAO engagement management template seems to provide guidance on the use of such analysis.

Finally, the conclusions drawn from the facts and analyses are necessarily in part a product of judgement and intuition. This in an inevitable part of the technology assessment process, but the engagement management template does not provide much guidance on it.

Reviewing the product. The GAO process provides for two major reviews of a report—the internal review at Gate 5 and the external agency review at Gate 6. The biometrics report passed through both gates, but the nature of the review process was modified somewhat to accommodate the special needs of a technology assessment.

Internally, the report went through at least three complete drafts that were reviewed at various levels in the organization (July 22, August 14, and August

22). The last of these was the formal internal review draft, which was reviewed by senior management. Given the evolutionary nature of policy analysis, this iteration seems appropriate. We would anticipate that any technology assessment (not just this first one) would benefit from a similar iteration and top-level review. Time in the engagement process should be made available for it, and earlier high-level review, which allowed the assessment team greater time to address fundamental structural issues in the analysis, would probably be wise.

After clearance by GAO management, the report was sent to the two affected operating agencies (Department of State and the INS) and to 19 external reviewers for comment. The external reviewers spanned a broad spectrum of backgrounds: six public interest groups, two elements of the biometrics industry, six university or research organizations, and six federal agencies with expertise in biometrics.

The external review process had a significant impact on both the engagement process and the substance of the final report. In terms of the process, the review

- Generated hundred of comments. We understand that the number received was substantially greater than the number received on the typical GAO report, in part because usual GAO practice limits comments to the affected operating agencies. However, the volume of comments is not out of line when compared with comparable study processes, such as that conducted by the National Research Council.
- Delayed the production of the final report by as much as two weeks because of the work required to address the comments.
- Required the engagement team to develop a procedure for ensuring that the comments were properly addressed. The GAO engagement management process requires that the agency comment letter and GAO's evaluation of it be included with the report. This approach ensures that the reader of the report can judge whether GAO has responded adequately to the comment letter, but it is not a practical approach for hundreds of external comments. Organizations accustomed to dealing with external comments typically have a self-audited procedure to ensure that they are thoroughly and fairly considered; again, the NRC's report review process serves as an example. This element is not part of GAO's standard engagement management process simply because external review is not part of the typical GAO report.

Substantively, the comments affected every element of the report. The following observations summarize our evaluation of their implications for the analytic process:

- **Factfinding.** A large number of comments offered corrections on the technical details of the report. Incorporating these comments was essential to the technical credibility of the analysis, although the comments did not materially affect the substance of the report. Many of these technical comments came from industry sources and from experts in federal agencies (e.g., NIST, FBI).

- **Scenario analysis.** Several comments, particularly from university scholars and research institutes, challenged major assumptions underlying the scenario analyses. Of special importance were comments on the size of the databases to be managed and on the treatment of uncertainty in biometric technology (e.g., false match and non-match rates). These comments resulted in adding important points to the scenario analysis (e.g., on the significance of False Error Rates in comparing to large watch lists) but did not affect the overall conclusions of the report.
- **Drawing conclusions.** A few comments, many of them from public interest groups, challenged the conclusions of the report. The comments were usually valid; for example, some of them stressed the point that a larger risk analysis of the overall border control problem is needed to draw definitive conclusions on biometrics. At the same time, it is fair to say that some of the recommended changes would have tilted the report in the direction of one or another policy outcome favored by the commenter. Responding to these comments required GAO to judge both their substantive validity and their effect on the report's overall balance and objectivity. Several changes were made to the report as a result of these comments, and we believe that the engagement team did a conscientious job of striking the right balance. On the other hand, judgements of this sort are ones that GAO management may wish to review, and for which the standard engagement process does not make provision.

If GAO conducts additional technology assessments, it should consider creating a more formal process for handling external review comments. In addition, we believe that making provision for more interaction before drafting the report with technical specialists and analysts in the field would have cut down on the number of technical and analytic comments. This would have substantially reduced the workload of responding to external comments and would also reduce the risk (not present in this case, as it turned out) of making a major revision to the report at the end of the process. We note, however, that providing for such interaction given the time constraints on the assessment and the absence of a routine process for this purpose would have been very difficult.

3.2. The Role of the National Research Council

From the early stages of the current assessment, the GAO recognized that it needed to seek the knowledge and advice of outside experts and affected stakeholders. GAO turned to the NRC for assistance in this phase of the work.

The GAO recently instituted a standing contract with the Division of Earth and Life Studies at the National Research Council to provide technical assistance, principally by convening and running topical workshops. When first proposed, this arrangement was controversial within the Academy complex because the products of these workshops are not subject to the Academy's usual extensive review process and there was a fear expressed by some NRC staff that these products would be represented as Academy products, while in fact the only role played by the Academy was that of convener.

When the GAO began the current technology assessment, they approached the NRC to conduct a workshop in which experts would have an opportunity to exchange views on technologies for biometric identification in border control. The task of organizing this workshop fell to NRC staff officer Richard Campbell. Using standard NRC methods, he assembled a diverse list of 59 potential participants by consulting with thought leaders in the field, academics, and other NRC staff. While retaining responsibility for the final selection of participants, Campbell shared his list with GAO to get their reaction and advice.

During the course of developing plans, it became apparent that a single workshop would not adequately cover all of the issues that needed to be addressed. It was decided to run two workshops, the first focused mainly on the key technologies, the second with a broader policy focus, addressing issues of public acceptance, civil rights, and stakeholder perceptions. For continuity, a few technical experts from the first workshop also participated in the second. Campbell completed the entire process of developing the two workshops and finalizing participant lists in about a month.

The workshops were chaired by Campbell, except for brief periods during the second workshop when GAO staff chaired. We sent a simple mail survey to all of the participants in the two NRC workshops. Forty-five percent of participants in the first and thirty-five percent of those in the second responded. (*Editor's note:* The results were summarized in an attachment to the original report).

Given the rapid timetable on which these workshops were organized, and the fact that GAO and NRC had not done this before, the participant evaluations strike us as very positive. Participants in the first, more technical, workshop report they were well briefed on the motivation and objective of the workshop and that they believed that it met its objective. Responses were a bit more mixed for the second, more policy-focused workshop. In both workshops, most respondents indicated that they thought the right set of people attended, although one respondent from each workshop indicated that it would have been good to have senior government stakeholders present, and one respondent indicated that the first workshop could have benefited from more systems-oriented people.

4. The GAO and Possible Future Assessments

While the final product has limitations, overall the GAO did a very good job on this pilot assessment, given the tight timetable on which the work was conducted and the fact that GAO had to innovate as it became apparent that important elements of their standard engagement process are not well suited to the task of technology assessment.

Much of the credit for the success of the pilot assessment must go to the excellent and innovative leadership provided at GAO by Naba Barkakati and Nancy Kingsbury. If GAO were asked to do assessments on a more regular basis, the availability of high-quality leadership and analytical staff would become an issue. These topics are discussed at greater length in the sections below.

As is apparent from the discussion in Section 3, process issues also bear importantly on GAO's future ability to conduct technology assessments. GAO's engagement management process could be extended to provide guidance on how to address the policy–analytic needs of technology assessment. However, the difficulties entailed in doing this should not be minimized. The existing engagement management process is one of GAO's core competencies, and a radical departure from it will require GAO to undertake the difficult institutional task of creating a new one and managing the tensions inherent in running an organization with two necessarily different cultures.

4.1. Staff Capabilities

Despite a history of tight budgets and its reputation as an auditing organization, the GAO has been able to attract and retain a small high-quality group of experts with backgrounds in science and engineering. Most of these experts are located in the Center for Technology and Engineering, which is part of the Applied Research and Methods team, under the direction of Nancy Kingsbury (*Editor's note:* a figure in the original has been excluded in this appendix).

In the ordinary course to GAO business, staff from the Center for Technology and Engineering are assigned to provide technical support and evaluation as part of ongoing GAO audits when the subject matter includes technical issues.

As discussed above, there are important differences between providing technical support for audits and performed technically based policy analysis and synthesis. The latter requires a consideration of a much broader set of issues, an effective integration of technical and non-technical considerations, and the creative identification, exploration, and evaluation of alternative policy options, of which technical systems make up only a modest, if important, subset.

Staff in the Center for Technology and Engineering have had little or no prior experience in performing such policy analysis and limited experience in taking the lead in managing engagements. Thus, complicating the tight time line of the current assessment has been the staff's need to "learn on the job." The staff with whom we interacted are bright, technically knowledgeable, and fast learners, but the subtleties and tools of high-quality technically based policy analysis are not quickly mastered. If GAO is going to continue to perform technology assessments for the Congress, it will need to invest in upgrading the policy–analytic skill of current staff, and it will probably also need to add staff who combine appropriate technical and policy–analytic skills. Such additions will be needed both at a junior, and also at a more senior, leadership level.

4.2. *Future Role of the NRC*

Using the NRC to convene and conduct expert and stakeholder workshops is a sensible strategy. NRC program officers routinely perform such tasks as part of their regular work and are very accomplished at doing so. They have wide contacts in the technical community and can readily identify key experts and stakeholders. They have appropriate facilities and staff support. And because of the high reputation of the Academy complex, persons invited to participate are probably more likely to agree than if the invitation were to come from a consulting firm, or perhaps even from the GAO itself.

If, in the future, the GAO is asked to perform additional technology assessments, it would be wise to continue to use the NRC to develop and conduct workshops of technical experts and stakeholders. It would be best if this could be done with more lead time and in a way that allows the subsequent assessment to be more effectively shaped by the workshop deliberations. It would also make sense for the NRC to designate a standing liaison person to assure continuity in the relationship and facilitate institutional learning and memory.

It might also be possible to develop a more formal mechanism by which the NRC could facilitate using some or all of the participants in these workshops as reviewers of GAO technology assessment products. In the review of the current assessment, this was done in a limited way, but at such a late date that it was difficult to do more than correct factual errors and note limitations in the analysis.

Asking the NRC to directly assist the GAO in the analysis and other functions of performing an assessment is probably not a viable strategy. Because of concerns about maintaining its independence and its reputation for high quality, it is clear from conversations that we held with NRC staff that the NRC is not willing to become involved as a collaborator in the production of products which are not produced according to its normal procedures and subjected to its own review and evaluation processes.

When the Congress wants the NRC to conduct a conventional NRC panel study on some topic, it has well-established mechanisms to ask for that directly and has no need to use the GAO as an intermediary.

4.3. *Issues of Scale-Up and Alternative Models*

If the Congress asks the GAO to perform more than two or three assessments per year, the GAO will face significant staffing problems. There are basically two options for scaling up the assessment capability: (1) add staff and build a special unit within the GAO that is dedicated to performing technology assessments; (2) perform assessments using carefully selected outside contract groups, under supervision by a much smaller special unit within the GAO, which is responsible for assessment oversight, quality control, and Congressional liaison. Which of these options GAO chooses would have important implications for the kind of in-house staff capabilities that they would need to develop. Option 1 would require the addition of staff with considerable policy–analytic skill and experience. Option 2 would require less of this, but still enough to provide informed guidance and supervision. Option 2 would also require the addition of staff with strong skills in project management and oversight.

Both options have advantages and disadvantages. First, there is the issue of managing institutional cultures. As noted above, the culture of the GAO is that of an audit organization. This is quite different from the culture required in an effective technology assessment organization. Either model would face the problem of developing and nurturing a group within the GAO which had a culture that was quite different from the parent organization. One could easily imagine staff tensions ("how come they get to ...") and other difficulties developing. As outsiders, we do not understand the internal operations of the GAO well enough to make a firm recommendation on this matter, but if the Congress decides that it wants to expand significantly the technology assessment capabilities of the GAO, this "two-cultures" issue should be carefully explored by both Congress and senior GAO administrators.

A model that involves substantially expanded external contracting has advantages, but also poses several challenges.

Advantages

- Can involve many more kinds of expertise and a greater depth of expertise than is ever likely to be possible with a dedicated GAO staff.
- Can more effectively build on existing work of an outside group and thus reduce the cost of some assessments.
- Can be readily ramped up and down in size depending upon the level of demand for analysis and available budget.

Challenges

- More difficult to keep the analysis group sensitive to and in tune with the needs of the Congress.
- Does not develop as large a permanent pool of technical expertise on the Hill to which Congress and staff can turn for informal advice when needed.
- May be more susceptible to budget swings.
- May be susceptible to Congressional pressures to distribute analysis to groups on the basis of political as opposed to quality considerations.

If GAO were to adopt a contracting model, we recommend that they limit the contracts to nonpartisan, nonprofit analysis groups in think tanks and universities, that they periodically compete several standing contracts with such groups, and that they require groups to participate in a program to sensitize them to the special nature and needs of Congress before they be approved to conduct studies.

Contracted studies would have to be carefully overseen by GAO staff. The quality and skills of a small permanent GAO staff would be a key factor in the successful operation of a contract model. They must be accomplished professionals who understand the issues of policy analysis, have good taste and high quality standards, which allow them to identify good and bad analysis, and broad experience that allows them to make informed choices among analysis groups. They must be sensitive to the needs and constraints of working for the Congress and effectively communicate these through firm and effective guidance to the study teams. They must supply or obtain the important institutional memory of how Congress has dealt with relevant issues in the past. They would need powers of persuasion and diplomatic skills so as to be able to negotiate appropriate study definitions, and when more than one group is involved in a study, assure harmonious collaboration.

Given that each has advantages and disadvantages, it is not clear to us whether the in-house or contracting model is superior. If GAO and the Congress wished to gather more experience before deciding, it could perform a small number of studies in each way and then compare the resulting products and processes.

4.4. The Need for Bipartisan Bicameral Oversight

If the Congress is going to make regular use of the GAO as a vehicle for technology assessment, it should give some thought to how to provide bipartisan bicameral oversight. Such oversight is important to assure that the assessment function

- remains responsive to the needs of the Congress;
- is not unduly influenced by the preferences and agendas of particular members, political groups, or parties;
- maintains a strict neutrality on issues of public policy; and
- uses scarce resources on problems that are of the greatest importance to the Congress and the nation.

Such bipartisan bicameral oversight is especially important in connection with the process of topic selection. Both the problems that are chosen for study and the way in which questions are framed are potentially very politically sensitive. If technology assessment activities at GAO are to avoid difficulties in the future, they will need some help in managing these risks. Realistically, that can only come through some form of bipartisan bicameral oversight.

Summary and Conclusions

Given the relatively short time that was available to perform the assessment and the fact that GAO has not previously engaged in such analysis, they have done a very good job. They successfully identified, described, and evaluated a range of relevant technologies. With assistance from the NRC, they also engaged, and in their report have reflected, the views held by experts, affected parties, and concerned stakeholders so that Congressional Staff and Members can accurately identify and weigh the trade-offs involved in policy choices.

This first GAO effort has been rather less successful in framing the analysis broadly in such a way as to address the full range of issues which the Congress is likely to have to consider and in providing analytically informed input that will support the needs of Congressional staff and Members as they refine and tune legislative products. These limitations result from the fact that while GAO has staff with excellent technical credentials, they have relatively little previous experience in framing and performing policy analysis. In addition, current internal GAO guidelines are not entirely consistent with the needs of technology-focused policy analysis and assessment.

If Congress concludes that it needs expanded technology assessment and related analytical capability to fill the gap between the short-term services provided by the CRS and the large-scale long-term studies of the NRC, it would be well advised to consider a range of alternative institutional mechanisms, of which studies conducted through the GAO are just one possibility. The problem should not be viewed in terms of choosing just one mechanism. It might be best to pursue several different strategies in parallel. The key point is to find better ways for Congress to obtain the types of assessment and analysis it needs in a timely manner so that it can be better informed as it addresses complex problems in technology and public policy.

While it will pose a significant challenge, we believe that if the Congress chooses to ask the GAO to perform additional technology assessments in the future, the problems we have identified can be resolved if GAO acquires more policy–analytic staff capabilities, makes appropriate changes in internal guidance and administrative arrangements, and makes effective use of outside expertise and contractors.

Reference

Fri, Robert, M. Granger Morgan, and William A. (Skip) Stiles Jr. 2002. An External Evaluation of GAO's Assessment of Technologies for Border Control, Washington, DC: United States General Accounting Office.

General Accounting Office (GAO). 2002. *Technology Assessment: Using Biometrics for Border Security.* GAO–03–174. Washington, DC: General Accounting Office.

Index

AAAS. *See* American Association for the Advancement of Science (AAAS)
Adversarial process of decisionmaking, limits to, 8–13, 173
Advice, technical. *See* Criteria; Technology assessment
Advisory models, comparison of, 159–60, 165
Advisory tradition in government, 13–15
  See also Committees, advisory; Congress, advisory agencies used by
Allocation of resources, 102–3, 104
  priorities, 155, 167
  See also Funding
Alternative models. *See* Institutions
American Association for the Advancement of Science (AAAS), policy fellowship programs, 134–35, 136–37, 139–41
Analysis. *See* Policy analysis; Technology assessment
Applied Research and Methods team at GAO, 218–19
Assessment. *See* Policy analysis; Technology assessment
Attributes. *See* Criteria; Institutions
Authority
  of institutions, 157–60
  of OTA, 186–87

Balance of power between Congress and executive branch in ability to analyze technical issues, 14–15, 57–58, 73–74, 137
Balanced analysis. *See* Objectivity
Basic Research and National Goals (NAS–House subcommittee study), 45–46(n.34)
Bias. *See* Objectivity
Bingaman, Jeff, Senator, and reconstructing a technology assessment capability within Congress, 113, 130, 210
Biometric technologies in border control, GAO pilot technology assessment, 130–31, 210–11, 212–16
Bipartisan and bicameral boards and committees, 102–3, 177, 178, 179
  for a dedicated organization, 161, 168
  for a distributed organization, 146, 148, 149, 152–53, 155–56
  for an expanded GAO, 226

Bipartisan and bicameral boards and committees—*continued*
at OTA, 58, 185–86
See also Oversight; Politics
Border control technologies, GAO pilot technology assessment, 130–31, 210–11, 212–16
Border Security Act (2002), and biometric technologies, 211
Boundary organizations (science and policy), 82–83
Budget. *See* Funding

Carnegie Commission on Science, Technology, and Government, reports on improving interaction between experts and Congress, 41–42
CBO. *See* Congressional Budget Office (CBO)
Collaboration
of analysts and stakeholders, 81
of GAO and NRC, 130–31, 212–13, 224
Committee on Science, Engineering and Public Policy, 36–37
Committees, 33
advisory, 13–14, 30–32, 147–48
House Subcommittee on Science, Research, and Development, 10–11, 37, 45–46(n.34), 55
at NRC, 120–22
Federal Advisory Committee Act Amendments of 1997, 205–7
See also Bipartisan and bicameral boards and committees
Communicating with Congress, 15–16, 41–42
distributed organization interactions, 153–54
OTA, 58–59, 81, 82
Confidentiality of CRS, 109
Congress, 14–15
advisory agencies used by, 32–40
expanding existing agencies or programs, 106–16, 139–42, 160, 165, 176, 177, 179–80
oversight of, 43(n.9)
analytic ability compared with executive branch, 14–15, 57–58, 73–74, 137
criteria for useful advice to, 26–28, 60, 63, 159, 174–75, 212–16
Congress—*continued*
expanding legislators' participation in Congressional S&E Fellowship Program, 141–42
increase in legislative activities related to science and technology, 55–57
need for information, 5–8, 25–28, 41, 53–58, 73–74
staff, 16, 28–32
Congressional S&E Fellows as, 30, 135–36, 143(n.1)
for a dedicated organization, 159, 162
for a distributed organization, 146, 148, 152–53
experts as, 43–44(n.14)
interaction with OTA staff, 86
rating of OTA reports, 66–67
size of, 43(n.8)
See also Institutions; Science policy; *specific agencies or programs by name*
Congressional Budget Office (CBO)
description, 35, 110–11
as possible technology assessment agency, 114, 116
Congressional Research Service (CRS)
description, 107–9
and a distributed policy analysis organization, 153–54
and OTA, 189
as possible technology assessment agency, 112–13, 115
and science policy, 32–34, 72, 174
Congressional Science and Engineering (S&E) Fellowship Program, 16, 134–35, 142–43
expanding the program, 139–42, 177
fellows as Congressional staff, 30, 135–36
history of, 136–39
Congressional Science and Technology Office, proposed model for U.S., 96–97
Congressional support agencies, 32–36
Consensus required for NRC recommendations, 122
Contract with America, and defunding of OTA, 36, 40
Contracting out
considerations for, 104, 146–48, 224–26

Contracting out—*continued*
European PTA agencies, 93, 94, 95, 96
to independent analysis group(s), 164–66, 167–68
proposals for U.S., 97
Cost
analysis by GAO, 213, 214, 215
of NRC studies, 121, 125
savings by OTA, 68–69
Credibility
as criterion for good policy analysis, 27, 81
of NRC, 119–20
of OTA, 86(n.4)
of policy institutions, 82, 112, 157–58, 159, 165
Criteria
for good assessments, 80–81, 147
for institutional success, 101–5, 111–12, 157–59, 160–61
rating of existing agencies, 114–16
for prioritizing resource allocation, 155
for useful advice to Congress, 26–28, 60, 63, 159, 174–75, 212–16
See also Science policy; *specific desired attributes by name*
CRS. *See* Congressional Research Service (CRS)

Daddario, Emilio
on NAS and Congress, 43–44(n.34)
on need of Congress for independent technical information, 57
as OTA director, 38, 55, 58–60
on Science Policy and Research Division, 44(n.21)
Danish Board of Technology, 92–93
Decisionmaking
and limits to adversarial process, 8–13
need for informed analysis, 4–8
See also Information; Policy analysis
Dedicated organizations, scenarios for, 157–63, 164–70, 177, 178
Design attributes for alternative institutional models, 82–86, 101–5
Direct contact model, weaknesses for science policy, 83–84, 162–63
Disinterestedness. *See* Objectivity
Distributed organizations, 84
compared with dedicated, 160
Distributed organizations—*continued*
European Parliament, STOA, 94, 96
scenario for an institute within Congress, 145–56, 177
See also Comparison of advisory models
Duplication of work, criticism of OTA, 72–73

Engineering Fellowship Program. *See* Congressional Science and Engineering (S&E) Fellowship Program
Engineers. *See* Experts
Environmental assessment, global, 81
European Parliament, Scientific and Technological Options Assessment (STOA), 94
as model for U.S., 91, 96
European parliamentary technology assessment (PTA) agencies, 90–97, 180–81
Executive branch
advisory committees, 13–15
analysis of technical issues, 12–13, 57–58, 73–74, 137
fellowship programs, 140–41
Expertise in governance
vs. political ideology, 6
views on, 3–5, 57–58
Experts
access to, 104, 158, 178–79, 180
as Congressional staff, 30–31, 43–44(n.14)
convened by NRC, 120, 126, 212, 224
independence of, 80, 165
at Institute of Medicine, 124–25
interaction with Congress, 41–42
weaknesses of direct contact with Congress model, 83–84, 162–63
See also Congressional Science and Engineering (S&E) Fellowship Program; Scientists; Staff

Fact-finding vs. scenario analysis at GAO, 221
Fainberg, Anthony, on Congressional S&E Fellowship Program, 135, 143
Federal Advisory Committee Act Amendments of 1997, Section 15, 130, 205–7
Federally funded research and development center contracting mechanism, 164–65

Fellowship programs
Congressional S&E Fellowship Program, 16, 134–43
executive branch, 140–41
improving use of alumni, 140
France, Office Parlementaire d'Evaluation des Choix Scientifiques et Technologiques, 92
incompatible model for U.S., 96
Fuel economy standards, comparison of NRC and OTA reports, 124
Funding
as constraint to institutional change, 175, 179
for a distributed organization, 146, 148, 149, 152, 153–54, 155–56
for an independent analysis group, 165–66, 168–69
NRC, 122–23, 128–29
OTA, 65, 70–71, 190
See also Allocation of resources

GAO. *See* General Accounting Office (GAO)
General Accounting Office (GAO)
description, 109–10
and NRC, 130–31, 212–13, 221–22, 224
and OTA, 189
pilot technology assessment, 130–31, 208–9, 222–23, 226–27
assessment of the assessment, 212–16
commissioning assessment and defining topic, 210–11
evaluation of, 208–27
reports, review process for, 219–221
technology assessment performance process, 217–22
and science policy, 34–35, 72
as a technology assessment agency, 113–14, 115–16, 179–80, 222–27
Germany, Office of Technology Assessment, 95
as model for U.S., 91
Gibbons, John
Clinton appointment, 46(n.42)
as OTA director, 38–39, 60–61, 64, 65
Global environmental assessments, 80, 81
Governance, view on role of expertise in, 3–5, 70

Health Effects Institute, 82
Hearings as information source, 5–6
Herdman, Roger, OTA director, 65
House Subcommittee on Science, Research, and Development, 10–11, 37, 45–46(n.34), 55
Huddle, Franklin P., CRS specialist, 33

Independent analysis group as policy institution, 169–70
funding, 168–69
mission, 166–67
organization and structure, 163–66
selection and operation, 167–68
Information
Congressional needs for, 5–8, 25–28, 41, 53–58, 73–74
vs. knowledge, 7–8, 23–24
See also Policy analysis; Research products; Technology assessment
Institute of Medicine, 191, 192
hiring experts for new studies, 124–25
studies mandated by Congress, 119
See also National Academies
Institutions, 121, 173–75
advisory agencies used by Congress, 32–40
Congressional oversight of, 43(n.9)
expanding existing programs, 106–16, 139–42, 160, 161, 176, 177, 179–80, 222–27
attributes for success, 101–5, 111–12, 114–16, 157–59, 160–61
direct contact model (private advice), 83–84, 162–63
institutionalizing technology assessment, 77–86
names for a new institution, 96–97, 179
structure
comparison of models, 159–60, 165, 176–78
of a dedicated organization in Congress, 157–63, 164–70
design attributes for alternative models, 82–86, 101–5
of a distributed organization, 84, 94, 96, 145–56
European PTA agencies, 90–97, 180–81
independent analysis group operated by NGO, 164–70

Institutions—*continued*
whether new ones needed, 16–18, 23–25, 41–42
See also specific agencies or programs by name
Interactive technology assessment, 78–79
Internet policy, disagreement over priorities, 20–21

Joint Committee on Atomic Energy, use of experts, 30–31

Learning
from assessments, 79, 80–81
promotion of, 86
Legislative activities related to science and technology, increase in, 55–57
Legislative effects of OTA analysis, 68*t*, 66
Legislative Reference Service. *See* Congressional Research Service (CRS)
Library of Congress
Division of Science and Technology, 32
See also Congressional Research Service (CRS)

Models. *See* Institutions

NAS (National Academy of Sciences). *See* National Academies; National Research Council (NRC)
National Academies
description, 118, 191–92
Federal Advisory Committee Act Amendments of 1997, 205–7
history of, 45(n.33)
role in technical advice, 36–37, 45–46(n.34), 73
studies mandated by Congress, 132(n.6)
list of mandates during the 102nd through 106th Congresses (1991–2000), 192–205
See also National Research Council (NRC); *specific agencies or programs by name*
National Academy of Sciences (NAS). *See* National Academies; National Research Council (NRC)
National Research Council (NRC)
current relationship with Congress, 118–25

National Research Council (NRC)—*continued*
funding mechanisms, 128–29
and GAO, 130–31, 212–13, 221–22, 224
history of, 13
process compared with OTA's, 123–25, 132(n.4), 132(n.5)
proposals for increased use by Congress, 125–28, 131, 176, 180
compared with dedicated organization, 160, 165
research products, 120, 125–28
See also National Academies; *specific agencies or programs by name*
National Science Foundation
and OTA, 189–90
and science policy, 73
Netherlands Office for Technology Assessment. *See* Rathenau Institute, Netherlands
Neutrality. *See* Objectivity
Nongovernmental organizations as hosts for independent analysis group, 164–70
NRC. *See* National Research Council (NRC)

Objectivity
as criterion for good policy analysis, 27, 63, 112, 158–59
how defined and achieved, 84
of NRC study process, 120
vs. relevance, 80–81, 169
in technology assessment, 102, 103, 105, 158–59
Office for Technology Assessment, Netherlands. *See* Rathenau Institute, Netherlands (RI)
Office of Technology Assessment (OTA), 173–74
accomplishments, 65–70, 107
communication with Congress, 81, 82
credibility of, 86(n.4)
demise of, 35–36, 39–40, 65, 70–74
early years, 58–60
Gibbons era, 60–65
insights from assessment experiences, 77–79, 81–86
as model for European PTA agencies, 90–91
origins of, 53–58, 107

Office of Technology Assessment (OTA)—
continued
origins of—continued
CRS role in, 108
Technology Assessment Act (1972), 184–90
"OTA process", 61–62
process compared with NRC's, 123–25, 132(n.4), 132(n.5)
rating of attributes essential for success, 114–16
role in technical advice, 37–41, 106–7
structure seen as best mechanism, 163
Technology Assessment Advisory Council, 58–59, 187–88
Technology Assessment Board, 58–59, 61–62, 102, 185–86
work continued by National Academies, 119, 131
Office of Technology Assessment, Germany. *See* Germany, Office of Technology Assessment
Office Parlementaire d'Evaluation des Choix Scientifiques et Technologiques, 92
incompatible model for U.S., 96
Operations research and policy analysis, 12–13
Organizational structure. *See* Institutions, structure
OTA. *See* Office of Technology Assessment (OTA)
Oversight
activities of Congress, 43(n.9)
of a dedicated organization, 158–59
of an expanded GAO, 226
of NRC, 122–23
of outside organizations, 146–48, 167–68
See also Bipartisan and bicameral boards and committees

Panel on Science and Technology, private meetings, 31–32
Parliamentary Office of Science and Technology, United Kingdom, 94–95
as model for U.S., 91, 96
Participation. *See* Public participation in technology assessment
Patriot Act (2001), and border control, 211
Policy analysis
boundary work, 82–83
democratic vs. technocratic views, 70
by distributed contractors, 145–49, 152–56
limits of fellowship programs, 138–39
at OTA, 62–64, 82–84
by outside organizations, 167–68, 178–79, 180
quality of, 15–16, 20(n.2), 39–40
criteria for Congressional advice, 26–28, 60, 63–64, 101–2, 174–75, 212–16
suggesting range of options, 101–2, 148, 150–51, 169
value in Congressional decisionmaking, 11–13
See also Information; Research products; Science policy; Technology assessment; *specific agencies or programs by name*
Policy options for institutional design, 101–5
Politics
controversy avoided by European PTAs, 96
and expert analysis, 6, 7–10, 14–15, 111–12
and funding research, 129, 149
and institutional design, 102–4, 165
and NRC support, 123, 132(n.6)
and OTA, 35–36, 38–40, 58, 61, 64–65, 70–71, 74
See also Bipartisan and bicameral boards and committees
President's Office of Science and Technology Policy, description, 72
Primack, Joel R., and Congressional S&E Fellowship Program, 136
Private advice. *See* Direct contact model
Process
and assessment effectiveness, 81, 85, 101–5
GAO
commissioning and defining biometric technology assessment, 210–11
performance of biometric technology assessment, 217–22
OTA, 61–62, 86(n.4)

Process—*continued*
OTA's compared with NRC's, 123–25, 132(n.4), 132(n.5)
stakeholder-based, 9–10, 86(n.4)
transparency of, 105
Products. *See* Research products
PTA. *See* European parliamentary technology assessment (PTA) agencies
Public participation in technology assessment, 78–79, 85
Danish Board of Technology methodologies, 93
OTA Technology Assessment Advisory Council, 58–59, 187–88
policy institutions stimulating public dialog, 69–70, 107, 159, 179
See also Stakeholder-based processes

Quality control. *See* Oversight
Quality of policy analysis. *See* Policy analysis, quality of
Quantitative analysis for decisionmaking, 11, 12–13

Rathenau Institute, Netherlands, 93–94
Reader-friendliness as criterion for good policy analysis, 63, 85
Relevance
as criterion for good policy analysis, 26–27
ensuring in a dedicated organization, 158–59
vs. objectivity, 80–81, 169
Reports. *See* Research products
Representative government and expertise, 3–5
Research and development (R&D), government expenditures increased, 54
Research products
attributes desired, 104–5
CRS, 108, 109
of a distributed organization, 148, 149
life cycle of a typical study by an outside organization, 149*f*
mandated by Congress for National Academies, 119, 132(n.6), 192–205
NRC, 120, 125–28
OTA, Congressional staff ratings of reports, 66–67
Research products—*continued*
of outside organizations, limitations, 174–75
See also Policy analysis
Resource allocation. *See* Allocation of resources
Review process for GAO reports, 219–21
Risk assessment and management, science advisory committees for, 13–14

S&E. *See* Congressional Science and Engineering (S&E) Fellowship Program
Scenario analysis in GAO pilot technology assessment, 219, 221
Science
increase in legislative activities related to, 55–57
questioning of its role in society, 54–55
Science advisory committees. *See* Committees, advisory
Science and Engineering Fellowship Program. *See* Congressional Science and Engineering (S&E) Fellowship Program
Science policy
changing nature of, 33
and Congressional S&E Fellowship Program, 137–39, 141–42
scientific advice in Congressional context, 23–28, 84–86, 106–7
weaknesses of direct contact with Congress model, 83–84, 162–63
See also Policy analysis; Technology assessment; *specific agencies or programs by name*
Science Policy Research Division (of Legislative Reference Service), 32, 107–8
contributions of, 44(n.21)
Scientific and Technological Options Assessment, European Parliament (STOA), 94
as model for U.S., 91
Scientists
and public participation, 36, 41–42, 54–55
See also Experts
Scribner, Richard, on Congressional S&E Fellowship Program, 137

Selection of outside organizations for policy analysis, 146–47, 154, 156, 167–68
Small-scale technology assessment, 95, 96
Social Learning Group, 80–81
Socio-technical mapping, 79–80
Sponsorship
  of Congressional S&E Fellowship Program, 134–35, 136–37
  and NRC efficiency, 122–23
Staats, Elmer B., GAO comptroller, 34–35, 110
Staff
  CRS, 108
  of a dedicated organization, 159, 162
  GAO, 110, 218–19, 223–24
  NRC, 122–23
  OTA, 62, 64, 65, 95, 186
  of outside organizations, 167–68
  See also Congress, staff; Experts
Stakeholder-based processes, 9–10, 86(n.4)
  See also Public participation in technology assessment
Standards. *See* Criteria
Stine, Jeffrey, on Congressional S&E Fellowship Program, 136, 137–38
STOA. *See* Scientific and Technological Options Assessment, European Parliament (STOA)
Strategic Defense Initiative, controversial OTA reports, 65
Studies. *See* Research products
Success. *See* Criteria

Technology
  increase in legislative activities related to, 55–57
  questioning of its role in society, 54–55
  See also Biometric technologies in border control
Technology advice. *See* Technology assessment
Technology assessment
  constructive, 79, 85–86
  in Europe, 90–97
  evaluation of GAO's first pilot, 208–27
Technology assessment—*continued*
  flexibility in, 96
  institutionalizing, 78–86
  interactive, 78–79
  real-time, 79–80, 85–86
  seeking and interpreting, 16
  topics requiring informed decision-making, *6b–7b*
  trends and obstacles in Congress, 23–42, 173–75
  of unintended consequences, 10–11
  See also Information; Institutions; Policy analysis; Research products; Science policy; *specific agencies or programs by name*
Technology Assessment Act (1972), 183–90
Technology Assessment Advisory Council, OTA public experts, 58–59, 187–88
Technology Assessment Board, Germany. *See* Germany, Office of Technology Assessment
Technology Assessment Board, OTA governing body, 58–59, 102, 185–86
  and "OTA process", 61–62
Timeliness
  and good policy analysis, 63, 85, 104–5, 161–62, 173–74
  of NRC studies, 121, 125, 126, 128
  of OTA studies, 65, 71–72, 104
Transparency of analytical process, 105

Uncertainty and scientific information, 12
Unintended consequences of policy decisions, technology assessment of, 10–11
United Kingdom, Parliamentary Office of Science and Technology, 94–95
  as model for U.S., 91, 96
Usability (reader-friendliness) as criterion for good policy analysis, 63, 85

Von Hippel, Frank, and Congressional S&E Fellowship Program, 136

Work products. *See* Research products